SAT Math: Keys and Tips

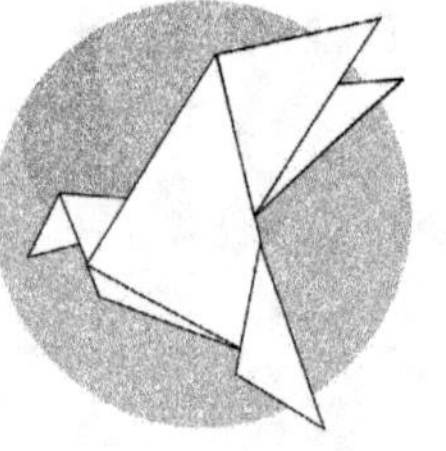

SAT:
MATH
Keys and Tips

Andy De Paoli

ISBN 979-1220071567

e-book

ISBN 979-1220071574

Firtst Edition 2020

© 2020, ADP. All rights reserved.

This book is dedicated to the memory of my sister, Aggie.

Contents

Notes   vi

1   Before we begin   1

   1.1   About Formulas   2

2   School Math   9

   2.1   Rationals   11

   2.2   Digits   14

   2.3   Long Division   15

   2.4   The Multiplication Table   16

   2.5   Exercises   20

3 More Ado About Numbers   **23**

    3.1 Properties 23

    3.2 Exponents 24

    3.3 Sequences 27

    3.4 Percentages 30

    3.5 Exercises 32

4 Algebra   **35**

    4.1 Operations 35

    4.2 Inequalities 38

    4.3 Functions 39

    4.4 Exercises 41

5 Linear Equations   **43**

    5.1 Angles 43

    5.2 Lines 44

    5.3 Exercises 49

6 Quadratic Equations   **51**

    6.1 Parabolas 51

    6.2 The Circle 60

6.3   Exercises 62

7 Growth, Decay   65

7.1   Exponential Functions 65

7.2   Exercises 69

8 Geometry   71

8.1   Plane Figures 71

8.2   Polygons 79

8.3   Circles (again) 83

8.4   Solids 85

8.5   Exercises 87

9 Trigonometry   89

9.1   Sin and Cos 89

9.2   Radians 91

9.3   Exercises 93

10 Complex Numbers   95

10.1 So-called Imaginary 95

10.2 Exercises 97

11 Word Problems   **99**

    11.1 Words into Equations 99

12 Permutations, Combinations   **103**

    12.1 Counting 103

    12.2 Probability 108

    12.3 Exercises 111

13 Statistics   **113**

    13.1 Average, Median, Mode 113

    13.2 Box and Whiskers 117

    13.3 Exercises 120

A Appendix   **123**

    A.1 Further Reading 123

B Appendix B   **127**

    B.1 A Bit About Number Systems 127

C Appendix C: Solutions   **135**

Preface

This book is not intended to replace the Official SAT Guide, but rather to complement it. The Official Guide was written by the people who actually write the tests and so I would dare say it is indispensable if you want to aim for a high score on the test. But the Guide's strength are the problems. It is not a math text book.

And math textbooks are designed to be used in a classroom setting and to be covered following a curriculum that lasts a year.

I needed something that could be used as a refresher course, and at same time aim specifically at the skills needed to do well on the SAT.

I have been looking for the best material to help prepare students for the SAT, but found that all had their good and their unsatisfactory aspects for my needs. I felt the need to have material I needed in one place, so I decided to write my own to fill the gap.

The purpose of this book is to help the students cut to the core of what the SAT is about: seeing the logic behind the problems, so that they may be broken down into easily solvable steps. In other words, use the key to unlock the problem.

1

Before we begin

If you've gotten this far in your studies you probably think you know how to study for an admissions math test. At least for school. But tests such as the SAT are not like your school math tests.

The SAT is not simply applying the formulas you learned in school. It is uncovering the **keys** that simplify the problems so that you can get to the solutions faster.

Simply doing one exercise after the other, checking answers and then proceeding on to the next one, is not the best way to improve your score. Simply trying for the answer is like shooting arrows at a target, one after the other, without learning the technique and building up the strength.

The reason you may not be reaching the score you are aiming for may be one or more of the following:

1. The underlying theory of the problem is not clear. If you got a wrong answer or even if you got it right but were uncertain why, you may need to review some long forgotten concept.

1

2. Your first instinct is to reach for the calculator instead of understanding the logic behind the problem. In other words, you don't step back from the problem and see if there is a twist in the question, so you start computing right away.

3. You were not able to find the right formula for the word problem, or not really understanding how to apply or manipulate the formula.

4. You just look at the numbers, instead of understanding what the question is really asking.

5. You work too fast, so your writing is disorganized and you make some silly mistake like wrong sign or inverting a ratio.

6. You didn't try quickly checking your answer on a difficult problem. If you answer is \$ 1,545 for a pencil maybe you missed a decimal point somewhere.

1.1   About Formulas

Let's talk about formulas for a moment. Those horrible things we had to learn long enough to use on the test and never really found much use for most of them after that. If you think about them as actually being shorthand notation that indicate the steps necessary to solve word problems, you may find them useful ()aside from the tests).

Let's look at the following formula:

$$2 \cdot n + k = l$$

It's not a terribly famous formula, in fact I just made it up. Would you know what it's for? Probably not. And even if you memorized it, but never really learned how it arose from a stated problem or didn't know what the variables mean, it would never be much help.

In standardized tests it is not just handy to know the formula. One must also know what the formula means and also know how to do what the formula tells us to do. Because formulas really are simply recipes for how to proceed to solve a problem.

Put the formula I wrote aside for a moment, and let's take a look at a few word problems.

Supposed you were asked one of the following questions:

- In a school yard there are 90 boy and girls, and there are 10 more girls than boys. How many girls are there?

- You have a piece of rope which you must cut into two pieces, one has to be 10 meters longer than the other. The length of the rope is 90 meters. How long will the longest piece be?

- You friend just bought two sweaters. The red one was 10 Euros more than the blue one. Together they cost 90 Euros. How much was the red sweater?

These, by the way, are based on real word problems that have appeared in the test.

Maybe you've noticed that they are actually the same problem clothed in different words. Maybe you've solved them right away. Even so, bear with me, I just want to make a point.

To solve them, let's represent 10 units (10 boys or girls, 10 meters or 10 euros) with small spheres.

We can set aside the last sphere, representing the ten greater, longer, or more expensive, and consider the remaining 80 (children, meters, or euros)

Now we divide the 80 in two parts, and join the 10 we had set aside to the right group.

I hope this made visualizing the problem and the solution, 50 of whatever, easier to see.

This was a logical approach that did not require knowing a formula. Many problems on the test can be solved by reasoning.

Or you can use another approach.

Transform the question into a formula. Take the formula I had given you, and let t represent the total (number, length, or amount). In other words, $t = 90$. And let k be the difference between the two desired values. In this case 10. And let n be the shortest or smallest value which is unknown.

By substitution we have $2 \cdot n + 10 = 90$ and solving for n gives us the shorter or smaller quantity.

I hope that makes it clear how a formula can accomplish with a few symbols as much as a long winded wording and explanation.

Formulas are great, not only if you know how to use them, but if you remember them correctly.

Not like the scarecrow in the movie *the Wizard of Oz* who got the Pythagorean theorem wrong when he said: "The sum of the square roots of any two sides of an isosceles triangle is equal to the square root of the remaining side."

(Did you catch the mistake?)

✎ **The best way to remember a formula is to learn how to derive it and, even better, explain it to someone else.**

In fact explaining something to someone is the best way to learn. Repeat something you have studied to someone else, simplifying and giving examples.

One great advocate of this was Richard Feynman. If you don't know who he was you may enjoy a biography called *Surely You're Joking Mr. Feynman*, a very enjoyable reading, and you can look up the "Feynman technique" on the Internet. You'll find tons of videos and explanations.

Plan Your Schedule

Regardless of how many days, weeks, or months you have before the test, set up a study routine. It is important that you set yourself a workable plan of action.

Determine how much time you can realistically dedicate to studying, whether one hour, two hours a day, or maybe only three times a week. Make a time sheet to log how much you've done. Stick to the schedule. If you have more than one hour a day then break it up in half hour minutes sessions. In the beginning, don't worry about solving problems fast, but rather study how they are worded, how to see what is being asked and if there are other approaches to solving the problems.

Jot down your weak areas. If there is one type of problem that you consistently get wrong or can't figure out how to approach, maybe a rate problem, then review it, read up on what others say, do some research. When you are finished and think you understand it, explain it to someone.

Did I say that already? Well, I can not repeat that enough. Explaining something is the best way to learn.

I recommend you do the problems in the *SAT Official Guide*. It was written by the same organization that wrote the actual tests, so they will be better than the exercises in many other books.

The *SAT Official Guide* contains several practice tests. I would recommend you choose one test to begin with, and not worry about speed just yet. Concentrate on understanding how the problems are structured and note the type of problems you have difficulty with instead. Soon you will see that the similarities among many problems.

✎**In the beginning take your time to understand the type of problems that are in the SAT.**

Observe the answers, very often, especially when they are all similar, note the differences and use them as a key to what the answer they are looking for is.

Ugh! Math

Many people have come to hate math, maybe because of the way it's taught, maybe it's because in our culture it is considered distant and abstract. If you can overcome the fear and the dislike and see the problems as provocation, like Sudoku or crossword puzzles, maybe you can at least learn to appreciate the challenge and feel satisfied when you solve one.

People who like puzzles get better the more they do them. If you have never done a New York Times crossword puzzle take a look. You may be surprised that anyone can do them, and yet there are people who learn to how solve them.

After all, isn't that what sports and games are also about? Trying to improve, to achieve something you couldn't before, and to prove yourself better than you were before?

Are these tests useful? Do they really measure something about you? There is an ongoing debate about the utility of tests. Maybe they are just one more prerequisite to restrict the selection the schools have to make by some arbitrary means. It may be better than selecting students based on height, hair color, or number of siblings. But in the end it should not matter. In the meantime use what tools there are to get ahead.

Others will be taking the same test. But you are not there in a race with them, you are there to do the best you can. You are there to be better than before.

The tests are not impossible, they do not involve higher math, calculus or notions beyond what you acquired in high school, so if you set your mind to it you can get as good a score as anyone else. That is, if you make the basic notions your own and if you practice.

About This Text

The topics are in a progressive order, that is, topics will come after the notions they are based on have been covered. Sometimes, however, there will be repetition of concepts because it may be useful to review the sane idea in a different context.

I have included a few problems (and the solutions in the appendix at the end) but not many.

You can find plenty in other books, especially *the SAT Official guide*, on the tests with plenty of opportunity to practice and do exercises. The few I have included are mainly to strengthen the concepts or to illustrate the ideas presented here.

School Math

As I said in the last chapter, the way to approach a problem in the SAT is to find the method that will get to the solution fast (and usually with the least calculation). In other words find what I call the *keys* that will unlock what otherwise would be a tedious calculation.

In this and the following chapters I will indicate important tips with the symbol

In the math or quantitative portion of the test you will, of course, be working with numbers. You may be asked for an integer solution, or you may find that you need to convert a fraction to a real number. So you will need to know what the different types of numbers mean: integers, rationals, irrationals, and reals. If you want a slightly (only just) more technical explanation of what these sets of numbers are and why we distinguish them I have added an overview in the appendix.

The most basic and intuitive numbers are the ones children learn first when they learn to count $\{1, 2, 3, \ldots\}$, called the counting numbers or **natural** numbers. In order to permit all the possible additions and subtractions we introduce the negative numbers and the zero. This set may be represented as

$\{\ldots -4, -3, -2, -1, 0, 1, 2, 3, 4 \ldots\}$ which is the set of **integers**. Notice that it contains as a subset the natural numbers which are the positive integers.

If an integer y is equal to x multiplied by an integer n then we say that y is a **multiple** of x, and x is a **divisor** of y. The positive divisors of 12 are { 12, 6, 4, 3, 2, 1 }, we should also consider the negatives { -12, -6,-4, -3, -2, -1 }, they are possible solutions in some of the problems you will encounter. But some problems only require the prime divisors which are { 2, 3 }

A number that has no divisors except itself and 1 is called a prime number. It is not a bad idea to know and recognize the primes below one-hundred: 2, 3, 5, 7, 11, 13, 17, 19, 23, 29, 31, 37, 41, 43 ....

You are invited to write out the rest, up to one hundred.

✎ Note that 1 is not a prime and that 2 is the only even prime.

Every integer greater than 1 can be expressed as unique product of primes (up to the order).

If we included 1 as a prime we could write 6 as the product $3{\cdot}2{\cdot}1$ or $3{\cdot}2{\cdot}1{\cdot}1{\cdot}1$ or any number of ones as factors, and we would not have a unique factorization.

All integers that are divisible by 2 are **even** integers, the others are **odd**. We can always represent an even number as the product of 2 times some smaller number m, that is as $2m$, and an odd number as $2m + 1$. This representation can be useful, for example, to prove that an odd number times another odd number is always odd.

$$(2m + 1) \cdot (2n + 1) = 2mn + 2m + 2n + 1 = 2(mn + m + n) + 1$$

If x is a **divisor** of y then it will "go into" y an exact number of times, otherwise it will go into y at most q times and leave a remainder r (which, of course will be smaller than x).

Formally we can state this as:

If x and y are two positive integers we can always find q and r so that $y = qx + r$ where $0 \leq r < x$.

Obviously if $r = 0$ then x is a divisor of y, or, to put it another way, y is divisible by x. For example 3 is a divisor of 45 but not of 47 since it leaves a remainder of 2, in other words

$$47 = 15 \cdot 3 + 2$$

We can add and subtract integers, and we will always obtain other integers, but not all integers can be divided and still be integers, for example an integer divided by another that is not its divisor, or one divided by a larger number, for example $\frac{16}{5}$, $\frac{2}{3}$ or $\frac{7}{19}$.

Very formally we say the integers are closed for the operation of addition and its inverse, subtraction. This property, called closure, is not satisfied for subtraction in natural numbers, since, for example, $7 - 11 = -4$ which is not a natural number. Good to know.

2.1   Rationals

So we introduce all possible fractions, and this is the set of **rational** numbers.

All the operations of addition, subtraction, and multiplication yield other rationals with the exception of division by zero.

Never, ever, divide by zero. Not even implied, so given $y/(x - 3)$, x cannot equal 3. A zero denominator is forbidden!.

Note that the rationals contain the integers, since they may be written as the integer over 1, e.g. $\frac{14}{1}$ or $\frac{1{,}357}{1}$.

They are called rationals because they come from the ratio of two lengths. You may remember from school, that the ancient Greek mathematicians, the

Pythagoreans, thought you could take any two lengths and they could be measured using a common unit. This seemed reasonable and intuitive.

For example the ratio of the sides of a rectangle that is 2 meters wide and 3 meters long would be $2 : 3$, or we can say the width is $\frac{2}{3}$ the length. Or you could be more exact and measure in millimeters, and the ratio could be $21 : 32$. In other words, with the right unit of measure you cold find the exact ratio of two lengths.

This may seem self-evident, but it turned out to be false.

Much to the dismay of Greek mathematicians, they discovered that some ratios cannot be expressed as a rational number, that is a ratio of two integers.

For example. geometrically speaking, for a square whose sides are 1 meter no rational number exists to define the length of the diagonal.

In the figure below we have drawn a square whose sides are a unit length and an arc from the diagonal to the extended side. Imagine swinging it down to the x axis, the end point of the arc will fall on the number we call the square root of 2, which is not a rational number.

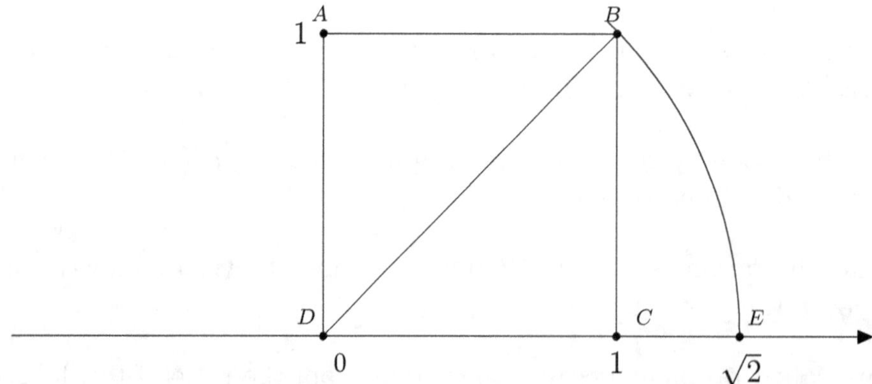

Algebraically we can state the same problem by saying that an equation such as $x^2 = 2$ has no solution among rational numbers. That is, there is no ratio $\frac{a}{b}$, where a, b are integers, that can satisfy the equation.

So we need to expand our system of numbers to include all the "irrationals", such as the solution to our equation, which we write as $\sqrt{2}$. All numbers that have no representation as rationals are called irrationals.

The new set, which includes irrationals and rationals, is the set of **real** numbers. We normally write reals as numbers with a decimal point.

Integers are the reals that have only zeros after the decimal point. Some rationals can be represented by a finite number of digits after the decimal, for example $\frac{1}{4} = 0.25$, but others cannot be represented exactly since the decimal part contains infinite digits, for example $\frac{1}{3} \approx 0.33333\ldots$, we normally write this repeating decimal with a bar of the repeating part, $0.\bar{3}$. More formally: rational numbers expressed in decimal notation (as reals) either terminate or repeat, as in the following examples:

$$\frac{9}{13} = 0.692307\overline{692307} \quad \text{or} \quad \frac{1}{8} = 0.125$$

But irrational numbers neither repeat nor terminate. For example:

$$\sqrt{2} = 1.41421356237309504880168872420969980\ldots\text{on to infinity}$$

$$\pi = 3.14159265358979323846264338327950288841\ldots\text{on to infinity}$$

For π you need only remember 3.14 if you need to do a calculation. Note that in almost all problems it is better not to solve for the value of π, or $\sqrt{2}$, or other irrationals, because the will either cancel out or actually be represented in the answer as the exact representation, π or $\sqrt{2}$. In other words replace them with the approximate numerical value only at the end if the question asks for it. You will see that this may only come up in problems that ask for an approximate answer.

2.2 Digits

Digits, by the way, are the ten symbols we use to represent numbers, that is the set $\{0, 1, 2, 3, 4, 5, 6, 7.8.9\}$ The number 274 is a three digit number since its takes three digits to represent it and the number $3,572.49$, for example, has six digits. The digit 5 is used to represent the hundreds and the digit 2 being the units digit (some books call it the ones digit).

We can imagine them as columns and so $3,572.49$ really is shorthand for $3000 + 500 + 70 + 2 + \frac{4}{10} + \frac{9}{100}$, and the position of each digit, that is, the column it is in, determines its real value.

3	Thousands
5	Hundreds
7	Tens
2	Units
.	(Decimal point)
4	Tenths
9	Hundredths

A problem may ask what the units digit will be for the number 13^{17}.

Obviously you won't be expected to calculate it, and most calculators will not even show it. So, what do we do?

Let's first notice that the units digit in a multiplication is only determined by the units digit of the multipliers, e.g. $743 \cdot 5327 = \text{xxxxxx}1$, so the first few powers of 13 will end in $13^2 \to 9, 13^3 = 7$ continuing $\to 1, 3, 9, 7, 1$ and so on. The pattern repeats itself and the seventeenth value will end in 3.

Another problem may require deciding what the value of some specific digits will be. For example what digits do A and B represent in the addition of the two digit numbers, AB and BA?

$$
\begin{array}{r}
A\,B \\
+\quad B\,A \\
\hline
1\,A\,1
\end{array}
$$

Since they are both two digit numbers neither A nor B is a zero. That means A + B must be greater than ten, giving $\{2, 9\}$ $\{3, 8\}$, $\{4, 7\}$ or $\{5, 6\}$ for A and B respectively as possible solutions. Note that all of them add up to 121, so the only one with A as the ten's digit in the result is A = 2 and B = 9.

2.3   Long Division

With real number division we obtain a real number as a result, which will consist of an integer part and a fractional (decimal) part, (either possibly zero). In other words, dividing y by x, we will obtain a real number z with an integer part q and a decimal r where $0 \le r < 1$, and we can write $y = qx + rx$.

For example $165/12 = 13$, with the remainder of 9, that is $165 = 13 \cdot 12 + 9$

$$
\begin{array}{r}
13 \\
12|\overline{165} \\
\underline{12} \\
45 \\
\underline{36} \\
9
\end{array}
$$

or we can continue, the division to obtain the decimals: $165/12 = 13.75$, and $165 = 13 \cdot 12 + 0.75 \cdot 12$

$$
\begin{array}{r}
13.75 \\
12\overline{)165.00} \\
\underline{12} \\
45 \\
\underline{36} \\
90 \\
\underline{84} \\
60 \\
\underline{60} \\
0
\end{array}
$$

Note that reminder is 9, which can also be found by taking 0.75 of the divisor, 12.

2.4   The Multiplication Table

There is no use knowing formulas if you stumble with basic operations. So make sure you know your multiplication table. If it takes you a few seconds to answer 7×8, 12×5 or $720 \div 8$ then it takes you a few seconds too many. It means you are wasting valuable time that could be dedicated more profitably during a test question.

If you hesitate with multiplication and division then, before you even begin doing exercises or reviewing your math equations, close your books. Don't go any further until you've spent some time re-memorizing you multiplication table. Maybe even pick a math exercise book and practice until you have total confidence and can beat your calculator.

✎Don't rely on a calculator. The calculations are almost all ones you can do in your head faster than typing it in a calculator, if you practice.

Another good thing to know are your perfect squares,

$$1, 4, 9, 16, 25, 36, 49, 64, 81, 100, 121, 144, 169, 196, 225 \ldots$$

You will find them useful in much of geometry: areas, distance on a coordinate plane and with the Pythagorean theorem, just to name a few.

Knowing your basic math is useful not just for the actual calculations you will encounter, but also for the rough approximations when checking your answers, just to see if you are in the ballpark. In fact, when doing exercises, practice rough approximations and become familiar with some of the techniques.

Once you have gained speed you can then learn many shortcuts that give you even more advantage when solving problems. You can find many books or websites with methods to speed up your calculations. You probably won't find them all useful or helpful, but perhaps you will find some congenial.

You know that in order to see if a number is even, or divisible by two, you simply need to look at the last digit.

Here are some quick tips to check for divisibility for single digits:

- Divisible by 2, even numbers, if the last digit is 0 or divisible by 2 (2,4,6,8)

- Divisible by 3, if the total of the digits added recursively is divisible by 3

- Divisible by 4, that is 2^2, if the last two digits are divisible by 4 (or last two digits 00), or divisible by 2 twice

- Divisible by 5, ends in 5 or 0

- Divisible by 6, if it is even and divisible by 3

- Divisible by 8, or 2^3, if the last three digits are divisible by 8 (or 000)

- Divisible by 9, that is 3^2, if the total of the digits added recursively is divisible by 9

Look at the list of the first few squares listed previously, notice that they only end in one of the following digits: $0, 1, 4, 5, 6, 9$. Perhaps numbers ending in 2, 3, 7 or 8 cannot be squares? Well that is the case.

You can prove it by noticing that the number minus the last digit ends in zero. If the last digit is d and the number minus that digit we can call n, the square of our number will then be $(n + d)^2 = n^2 + 2dn + d^2$.

We saw that all single digit numbers end in one of the above list, and d^2 is the square of a single digit number. But $(n^2 + 2dn)$ ends in zero and adding d^2 leaves the last, or units, digit unaltered so our squares must all end with one of the digits in the list.

If you know your perfect squares you won't have much trouble then remembering their square roots.

There are plenty of books with exercises that you can practice, and of course the more you do the more you will improve. But as you work through the problems you must think about what you are doing, and why. Especially at the beginning of those areas you are not familiar with.

If you got a problem wrong or you are unclear about the explanation it is important not only that you follow the explanation step by step until you are clear as to what you are supposed to do. Or better yet, see if you can find alternative approaches. These books do not always give the best tips, but only standard procedures.

This is especially important to remember if you get several similar problems wrong. Try to find a common approach. The essential part of working on math problems is to be flexible and find different ways of attacking a problem. For example geometry and algebra are two different fields and may look at the same problem in two different ways.

It is also good exercise for the gray matter of the brain.

Extra Material

If the test date is not too close, you may look up techniques for fast mental calculation. One such technique when multiplying a number by five, as an example, is to add a zero to the number and divide by 2. It is usually much faster than multiplying each digit by five, summing and carrying.

A very simple example: to multiply 126 by 5, just think 1260, and divide by two, which is 630. To divide by 5 a similar technique is to double the number and divide by ten (move the decimal point one digit to the left). For example, $\frac{16}{5} = 32/10 = 3.2$

Advanced Extra

But what of the square roots of numbers that are not squares? This is unlikely in tests such as SAT. But if you wish to learn a technique to find the square root of a number n (which is not a square) a decent approximation may be obtained as follows:

1. find the closest square smaller than n, say it is m and its root is r (that is your integer part),

2. write the difference between your number and the perfect square: $p = n - m$, and that will be the numerator,

3. and twice the square root r will be your denominator.

Let's try two examples:

You need to find $\sqrt{87}$. You know it which is just bigger than 81, which is 9^2. So the integral part is 9, that is, the root you are looking for will nine point something. The numerator will be $87 - 81 = 6$, and the denominator will be $9 \times 2 = 18$, so

$$\sqrt{87} \approx 9\frac{6}{18} = 9\frac{1}{3} \text{ approximately, the real number } 9.33$$

Compare this result that of a calculator: 9.38.

Another example:

You need to find $\sqrt{110}$. which is greater than 100, 10^2. The integral part is thus 10. The numerator is $110 - 100 = 10$, and the denominator will be 20, so

$$\sqrt{110} \approx 10\frac{10}{20} \text{ or approximately } 10.5$$

2.5   Exercises

Answers are in the back of the book, Appendix C.

1. Simplify
$$\sqrt{27}$$

2. Simplify
$$\sqrt{138}$$

3. Which of the following are divisible by 9, by 8?
$$\{117, 144, 152, 162, 192, 198, 904, 1017, 1024\}$$

4. Simplify

$$\frac{87,123}{113}$$

5. Find the divisors of 1,820. Which are prime factors?

3

More Ado About Numbers

3.1 Properties

Some properties of numbers you are probably familiar with, even if not expressed formally. The commutative property of addition and multiplication tells us that if you change the order the resulting sum or multiplication is the same: $a + b = b + a$ and $a \cdot b = b \cdot a$

The associative property also holds for addition and multiplication:

$$a + (b + c) = (a + b) + c$$

and

$$a \cdot (b \cdot c) = (a \cdot b) \cdot c$$

And you of course know that these properties do not hold for subtraction and division. $7 - 3 \neq 3 - 7$.

The third property which is very (I mean very) often necessary to invoke is the distributive property:

$$a(b + c) = ab + ac$$

It probably seems obvious expressed this way, but quite surprisingly many overlook it in expressions such as $x + 3.2x$, which can be written explicitly as $1 \cdot x + 3.2 \cdot x$, where we have evidenced the coefficient 1, and applying the property, regrouping terms, we get $x(1 + 3.2) = 4.2x$

Absolute Value

Absolute value, also called *magnitude*, or numerical value, can be defined as

$$|a| = \begin{cases} a & \text{if } a \geq 0 \\ -a & \text{if } a < 0 \end{cases}$$

This formal definition of absolute value simply means that a number is always positive, so $|-5| = 5$, $|5| = 5$, $|-345| = 345$.

If you are asked what the value inside the absolute value sign is you must remember that both the positive and negative values will give the positive result, in other words, given $|x| = 5$, x may be either 5 or -5.

Another example, since the expression $|2 \cdot x - 3| = 11$ means that both $2 \cdot x - 3$ and $-2 \cdot x + 3$ are equal to 11.

The same approach is used for inequalities. To solve for

$$|x + 7| \leq 9$$

we simply work both equations, $x + 7 \leq 9$ and $-(x + 7) \leq 9$, giving us $x \leq 2$ and $x \geq 16$, in other words $-16 \leq x \leq 2$.

3.2 Exponents

Just as multiplication is shorthand for repeated addition, $4+4+4+4+4 = 4 \times 5$, so exponentiation is shorthand for repeated multiplication. It is important then to be clear that $3^2 + 3^3 \neq 3^5$. In fact $3^2 + 3^3 = 3 \cdot 3 + 3 \cdot 3 \cdot 3$. You could factor out 3^2, and get $3^2(1 + 3) = 3^2(4)$.

You can combine terms when the expression is a product of two exponents with the same base. $5^3 \cdot 5^4 = 5^{3+4} = 5^7$ since $5^3 \cdot 5^4 = 5 \cdot 5 \cdot 5 \cdot 5 \cdot 5 \cdot 5 \cdot 5$.

A power raised to another power can be simplified by multiplying exponents. $(7^3)^4 = 7^{12}$.

A negative exponent means that the expression is the **reciprocal** or multiplicative inverse of the expression. That means that $3^{-2} = \frac{1}{3^2}$ and, for example,

$$\frac{1}{7^{-3}} = 7^3$$

A fractional exponent is a way of expressing roots and powers. A simple example, $64^{1/2} = \sqrt{64} = 8$ and $27^{1/3} = \sqrt[3]{27} = 3$.

Since positive exponents mean multiplication, negative exponents mean the inverse operation, which is division, so $a^{-m} = \frac{1}{a^m}$, and more generally

$$\left(\frac{a}{b}\right)^{-m} = \left(\frac{b}{a}\right)^m$$

The power as an integer tells us how many times the base is multiplied by itself, so the fractional exponent expresses the root of that number:

$$a^{\frac{1}{m}} = \sqrt[m]{a}$$

Putting these ideas all together we can simplify the expression $36^{-3/2}$

$$36^{-\frac{3}{2}} = \frac{1}{\left(\sqrt{36}\right)^3} = \frac{1}{6^3} = \frac{1}{216}$$

In general, the rules for exponents are the following.

The product of two powers with the same base is the base to the sum of the exponents. That's a mouthful but it simply means :

$$a^m \cdot a^b = a^{(m+n)}$$

This is easy to see in practice:

$$3^3 + 3^2 = (3 \cdot 3 \cdot 3) \cdot (3 \cdot 3) = 3 \cdot 3 \cdot 3 \cdot 3 \cdot 3 = 3^5$$

The other property just as easy to very with small numbers.

$$(a^m)^n = a^{(mn)}$$

Here are some numerical examples:

$$5^3 \cdot 5^2 = 5^5$$

$$\left(7^3\right)^2 = 7^6$$

$$12^{-2} = \frac{1}{12^2} = \frac{1}{144}$$

$$7^{\frac{1}{3}} = \sqrt[3]{7} = 3$$

$$64^{-\frac{2}{3}} = \frac{1}{64^{\frac{2}{3}}} = \frac{1}{\sqrt[3]{64^2}} = \frac{1}{(\sqrt[3]{64})^2} = \frac{1}{16}$$

and you probably remember that $17^0 = 1$. Why? Not just because your teacher told you, but because:

$$1 = \frac{a^m}{a^m} = a^m \cdot a^{-m} = a^{m-m} = a^0 \rightarrow 1$$

Two more properties you should know are that 1 raised to any power is still one, and similarly, zero is always zero when raised to any power.

$$1^n = 1 \text{ and } 0^0 = 0 \text{ for all } n$$

A square is always positive.

✎ **Even powers are always positive, and negative powers depend on whether the base is negative or positive.**

So if you were asked which is greater $A = (-7)^4$ or $B = (-7)^5$ it should be immediate that A is greater, without doing any calculations.

3.3 Sequences

A mathematical **sequence** is a string of numbers that follows a particular pattern. The individual elements are called **terms**.

An arithmetic sequence is obtained by adding a constant term, the simplest being a sequence of integers (we are adding one to each successive term), $\{1, 2, 3, 4, 5 \ldots\}$. Another sequence may be obtained by adding 3 (called the common difference) to the last term obtained, in an example starting with 9: $\{9, 12, 15, 18, 21, 24 \ldots\}$

If the first term is a, and we add a common term d to each, the sequence will be

$$a, a + d, a + 2d, a + 3d, \ldots, a + nd$$

The nth term will then be $a + nd$

It is definitely useful to know the formulas for sums of consecutive integers of n terms,

$$1 + 2 + 3 + 4 + \ldots + n = \frac{n(n + 1)}{2}$$

or more formally:

$$\sum_{x=1}^{n} x = \frac{n(n+1)}{2}$$

The sequence of odd numbers $\{1, 3, 5, 7, 9, 11 \ldots 2 \cdot -1\}$ we see that the sum of the first n odd numbers is the position of the nth number squared.

A geometric sequence is obtained by multiplying by the same value, for example, the common ratio 3 beginning with the number 6: $\{6, 18, 54, 162, 486 \ldots\}$

Similarly to the arithmetic sequence we can call the first term a, and each successive term multiplied by d, the sequence will then be

$$a, a \cdot d, a \cdot d^2, a \cdot d^3 \ldots \cdot a \cdot d^n$$

Note that there are $n + 1$ terms because we would be starting (formally) with zero: $a \cdot d^0 = a$

So, for example, in the sequence $5, 10, 20, 40, 80, 160$, the 6th term is $5 \cdot 2^{6-1} = 5 \cdot 2^5 = 5 \cdot 32 = 160$

The **range** is the maximum value minus the minimum. (max - min).

Whereas the number of integers from the minimum a to the maximum b inclusive is $b - a + 1$. For example in the sequence $\{3, 4, 5, 6, 7, 8, 9, 10, 11\}$ the range is 8, and there are $11 - 3 + 1 = 9$ integers. Another example, the number of integers between 15 and 42 is 28.

The range between n and m, $n < m$, is $m - n$, and there are $m - n + 1$ integers in the range.

Before tackling the more difficult problems in the test, try experimenting with short sequences that you can also verify quickly. For example, the sequence of integers between 3 and 8 is $\{3, 4, 5, 6, 7, 8\}$, and you can immediately see that the range (distance or length) is $8 - 3 = 5$, whereas the number of integers between 3 and 8 inclusive is $8 = 3 + 1 = 6$.

Or, another example, the sum of even numbers in the sequence $\{2, 4, 6, 8, 10\}$, Try to spot a pattern after adding the first few, $2, 2 + 4 = 6, 2 + 4 + 6 = 12$ and so on. Compare the sequence with that of the integers from 1 to 5, and you can see that it is always twice for the corresponding position. Of course that is because the sum $2 + 4 + 6 + 8 + 10 = 2(1 + 2 + 3 + 4 + 5)$.

Other sequences

A famous special sequence is the Fibonacci Sequence:

$$(0, 1, 1, 2, 3, 5, 8, 13, 21, 34, 55, 89 \ldots)$$

defined as

$$F_n = F_{n-1} + F_{n-2}$$

Which means we add the two previous numbers to obtain the new one and add that to the end of our list. We then proceed the same way with the last two numbers of our list.

There are other sequences which may be interesting in their own right, but I have not seen much of them in the tests.

One is that of the so called Triangle Numbers, obtained by adding increasing integers

$$(1, 3, 6, 10, 15, 21, 28, 36 \ldots)$$

and the Square numbers, obtained by adding increasing odd numbers:

$$(0 + 1 = 1, 1 + 3 = 4, 4 + 5 = 9, 9 + 7 = 16, 16 + 9 = 25, 25 + 11 = 36 \ldots)$$

3.4 Percentages

When given a problem involving percentages is very important to express the percentage as either a fraction or a decimal.

The word "percent" means "over one-hundred", so, for example:

$$23\% \rightarrow \frac{23}{100} = 0.23$$

When, in a word problem, expressing the percentage as the decimal equivalent, note that the decimal point comes between the tens and the hundreds digit. For example $327\% \rightarrow 3.27$, $15\% \rightarrow 0.15$, or $5\% \rightarrow 0.05$

In many problems you will be asked by what percentage something has changed, for example something that cost 2,500\$ now costs 2,900\$. What was the percentage increase?

The way to solve it is by taking the difference between the old and the new value, divided by the old value.

$$\frac{\text{new - old}}{\text{old}} = \frac{change}{old}$$

So the solution to the problem given would be

$$\frac{2900 - 2500}{2500} = \frac{400}{2500} = \frac{4}{25} = 0.16$$

The increase was therefore 16%.

Be careful of the trap in a question where the percent increase and the subsequent decrease, or vice-versa, the new value will **not** be the same. If 100 increases by 10% ($100 + 10 = 110$) and then decreases by 10% ($110 - 11 = 99$)

Some problems may be solved in one form much easier than in another. If you have trouble understanding that 75% is shorthand for 0.75 or $\frac{3}{4}$ then stop and practice some transformations.

For example 10% of 50 is just a matter of multiplying $50 \cdot 0.1 = 5$, while 60% of 25 can be readily solved as $3/5 \cdot 25 = 3 \cdot 5 = 15$.

The most common percentages in tests are multiples of 25 and multiples of 20.

To learn these without always calculating them over again, think of 100%, that is 1.0 as made of four blocks, each being 25% or 0.25, or $\frac{1}{4}$, this should help go quickly from one notation to the other.

25%	50%		$\frac{1}{4}$	$\frac{1}{2}$
75%	100%		$\frac{3}{4}$	$\frac{1}{1}$

You can also divide 100% into five blocks of 20% each:

$20\% = \frac{1}{5}$	$40\% = \frac{2}{5}$	$60\% = \frac{3}{5}$	$80\% = \frac{4}{5}$	$100\% = \frac{1}{1}$

This should help you visualize the decimals and the fractions. Once you visualize this you can also use it to approximate, for example, 38% is 0.38 of the given quantity, so it is a bit less than 0.40 or $\frac{2}{5}$. When you write down formulas involving percentages, do not write % (the percent symbol) but convert it to its decimal equivalent.

Note the relation between the percentage and the fraction, such as $1/20 = 0.05$ and $1/5 = 0.20$, or 0.25 and $1/4$, and $0.04 = 1/25$.

Ratios

There will be several problems in the test concerning ratios.

What are ratios? Generally they are the comparison of two quantities, that is one value relative to another. The number of boys in a class compared to the number of girls. Or it could be a mixture of several substances, or the ratio of sides of a triangle, and so on.

We can denote a ratio in several ways: $x : y$, or "x to y", or as a fraction x/y. You use ratios very often, even if you didn't realize it, when you talk about "kilometers per hour", you are using a ratio km/hr. When one mentions probability, such as four chances in a fifty-two to pick a certain value card from a deck. They are using a ratio

Given a ratio, just like a fraction, it is usually better to simplify them, so a ratio $24 : 36$ can treated as $2 : 3$. Both sides of the ratio can be multiplied or divided by the same amount.

If you are given three quantities compared to each other, $x : y : z$. For example $3x : 5y : 2z$, this is called a continued ratio, and they can be pairwise compared, so the ratio of x to z is $3x : 2z$

3.5   Exercises

Answers are in the back of the book, Appendix C.

1. Find the sum of integers from 9 to 75, where each element of the sequence is a multiple of three.

2. Simplify the expression

$$98^{\frac{1}{2}}$$

3. What is 15% of 130?

4

Algebra

4.1 Operations

When we have algebraic expressions with variables, we can perform the same operations as with numbers. The unknown quantities, called **variables** are symbolized by letters such as x, y, a, b, m, or n.

An expression containing variables is an **algebraic expression**, such as the following

$$8\frac{720d^2}{9d^2(-8d)}$$

and it is an **equation** when there is, of course, an equal sign

$$8\frac{720d^2}{9d^2(-8d)} = 1$$

An example of multiplication with one variable $3x \cdot 3x^2 = 5x^3$. Bot that we multiply the coefficients 3 and 2, and the powers with the same base, x, are added.

35

You are certainly familiar with the multiplication of binomials.

$$(ax + c)(bx + d)$$

You should be able to solve these very fast since they will be included in a large part of the math problems. Note that the coefficient of x^2 will be the product $a \cdot b$, and the constant term will be the product of the two constants, $c \cdot d$.

Sometimes, looking at the answers to choose from this may be enough to identify the correct one. If not, the middle term will be $(a \cdot d) + (c \cdot b)$, as in the example:

$$(5x + 2)(3x + 7) = 15x^2 + 35x + 6x + 14 = 15x^2 + 41x + 14$$

You should also be able to recognize the following products, and the inverse operation, factoring, for sums

$$(a + b)(a + b) = a^2 + 2ab + b^2$$

for a product of differences

$$(a - b)(a - b) = a^2 - 2ab + b^2$$

and the product of a sum and difference

$$(a + b)(a - b) = a^2 - b^2$$

Note in the last case that the middle term has canceled out. This last one is called the "difference of two squares", and there are many cases in which it is important to recognize since it will make factoring a lot easier.

For now a simple example is $16x^2 - 9y^2 = (4x + 3y) \cdot (4x - 3y)$

Again, just as with numbers, we can extract a common factor

$$4a^3 + 2a^2 + 6a = 2a \cdot (2a^2 + a + 3)$$

Since you are not expected to solve equations higher than quadratics, if the problem is a polynomial of third or higher power then you can probably factor out some power of x and reduce one of the terms it to a quadratic that is solvable.

We will look at some more of these concepts when we get to quadratics.

One manipulation you must be able to do quickly and correctly when there is a fraction on either side of an equation. In the example you can think of bringing the x up to the other side, but this is only a shorthand way of saying, multiplying by x on both sides. Similarly, bringing down a value to the other side is actually a division on both sides of that value.

$$\frac{8}{25} = \frac{2}{x} \rightarrow \frac{8}{25} \cdot \frac{1}{2} = \frac{2}{x} \cdot \frac{1}{2} \rightarrow \frac{4}{25} = \frac{1}{x} \rightarrow x = \cdot \frac{25}{4} = 6 \cdot \frac{1}{4}$$

An expression which is a sum of terms, each of which has a power of x times a number called a constant, is a **polynomial**. The following is a polynomial

$$12x^3 - 5x^2 + 6x + 7$$

with four terms, $12x^3$ is the first term, $-5x^2$ the second, $6x$, and the constant term is 7, which satisfies the definition since it may be written as $7x^0$, because $x^0 = 1$. The highest power of x in this polynomial is 3 so it is a third degree polynomial.

Generally we write the highest power of x first and then write them in decreasing order.

If the highest power is one, then it is a linear polynomial. Many problems you will encounter in the test will be of the second degree or **quadratic** polynomial.

We may find (especially in tests) that we are given an equation with an expression which we need to manipulate to put it in a polynomial form in order to solve the equation. An expression such as

$$\frac{7x^3 + 2}{5x - 13}$$

is not the sum of powers of x times a coefficient and is therefore not a polynomial.

Remember to simplify equations and expressions when possible. The expression

$$\frac{4x + 12}{6 + 8x^2}$$

can be simplified to

$$\frac{2x + 6}{3 + 4x^2}$$

Given the equation $24x^2 + 16x + 4 = 0$, divide both sides by 4 to obtain $6x^2 + 4x + 1 = 0$

✎ Be sure to see if there is a common divisor for all constants and coefficients which can help simplify the problem.

4.2   Inequalities

Let's look at inequalities a bit more closely. Remember that they can be treated like equalities when manipulating them with positive integers. That is, you can add, subtract, multiply or divide both sides by any positive integer and the inequality still holds.

So $2x \geq y \leftrightarrow 4x \geq 2y$ hold for the same values of x and y, or $3z + 5 > 14 \leftrightarrow 3x > 9 \leftrightarrow x > 3$.

✎ However, multiplying or dividing by a negative number reverses the inequality. If $x < 5$ the $-x > -5$.

We can verify with some values plugged in to the two inequalities.

x	$x < 5$	$-x > -5$
3	$3 < 5$	$-3 > -5$
0	$0 < 5$	$0 > -5$
-3	$-3 < 5$	$3 > -5$

If you are given a problem with more than one inequality, it is usually best two treat them separately. For example, given $3x > 6y > 2z$, you can simplify the two inequalities $x > 2y$ and $3y > z$.

Inequalities with absolute values, just as we saw previously, we can treat the positive and negative values of the expression inside the bars. So $|x + 3| \geq 15$, you would find the values of x for $x + 3 \geq 15$ and for $-(x + 3) \geq 15 \rightarrow$, simplifying we would have that it is true for all $x > 12$ and $x \leq 18$.

You are encouraged to verify the solution.

4.3 Functions

A function defines how we are to operate on one or more elements of a set to obtain elements of another set.

For example, the function defined as squaring a number means multiplying any number by itself and obtaining a positive number. Note that the initial value may be either negative or positive. This is expressed in shorthand as

$$f(x) = x^2$$

You can think of a function as black box that turns one value into another

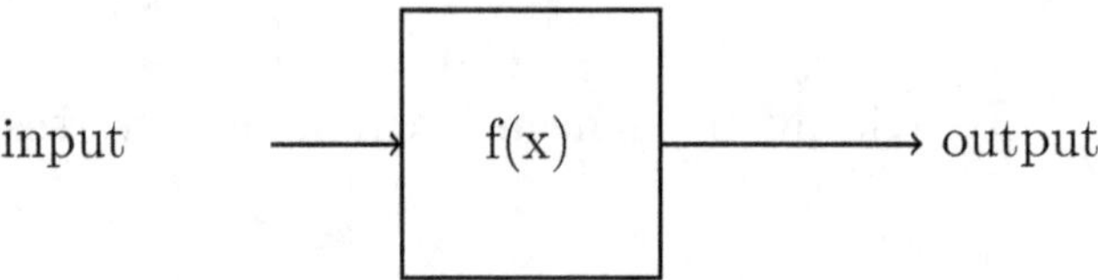

The notation $f(x)$, read as "f of x", is a definition of the function that tells you to substitute any occurrence of the x on the right side with any value specified. For example, given the function $f(x) = x^2 + 3x - 7$, to find the value of $f(5)$ we rewrite the definition, substituting 5 for the x

$$f(5) = 5^2 + 3 \cdot 5 - 7$$

In other words, the left side tells that the function varies as x varies. The right side tells you that whatever will replace the x in the parentheses will substitute the variable on the right side.

You can think of the x as an empty box to be filled in with the value given.

$$f(x) = \square^2 + 3\square - 7$$

The most common functions in the tests are linear functions,

$$f(y) = mx + b$$

and quadratic functions

$$f(x) = ax^2 + bx + c$$

The argument or input is x and the output (or the image) is usually represented by y, and so we denote the function as $y = f(x)$

The tests often employ notations other than $f(x)$, such as $g(x)$ or $h(x)$. Or even stranger ones such as $\lceil x$ or $\star x$ Don't let that fool you, they all mean the same as $f(x)$.

There are also compositions of functions. For example, $f(x) = 3x^2 + 2$, and $g(x) = x - 3$. Given the problem with the composition of functions, $f(g(7))$ we first solve for $g(7)$ and the output becomes the input for the out function $f(x)$,

$$f(g(7)) = f(7 - 3) = f(4) = 3 \cdot 4^2 + 2 = 50$$

Functions can even be nested in themselves. For example, for the function $f(x) = (\sqrt{x})^3$ find $f(f(81))$

$$f(f(8)) = f(9^3) = f(729) = (\sqrt{729})^3 = 27^3 = 19,683$$

Functions can be represented graphically, as we will see ion the sections on linear and quadratic equations. They may also be represented as tables, from which we sometimes have to derive the function.

X	$\rightarrow$	Y
0		2
1		5
2		8
4		14

Can you derive the function from the above table? For $x = 0$ we have $y = 2$, the y intercept is therefore 2. You may have noticed that it is a linear relationship, and in fact $y = 3 \cdot x + 2$

4.4 Exercises

Answers are in the back of the book, Appendix C.

1. Solve for $f(g(5))$

$$f(x) = x^3 + 3x$$
$$g(x) = 2x^2 - 5$$

2. Simplify

$$\frac{8x^2 - 50}{4x - 10}$$

5

Linear Equations

5.1 Angles

When two lines cross or line segments meet they define an angle. If they are perpendicular they then form a right angle, that is 90° when expressed in degrees, or $\frac{\pi}{2}$ when expressed in radians.

We will see more about angles in later chapters.

✎ A straight line is 180° or π radians. It helps to remember that π is the ratio of the circumference to the diameter.

When a line crosses two parallel lines the corresponding internal angles are equal.

43

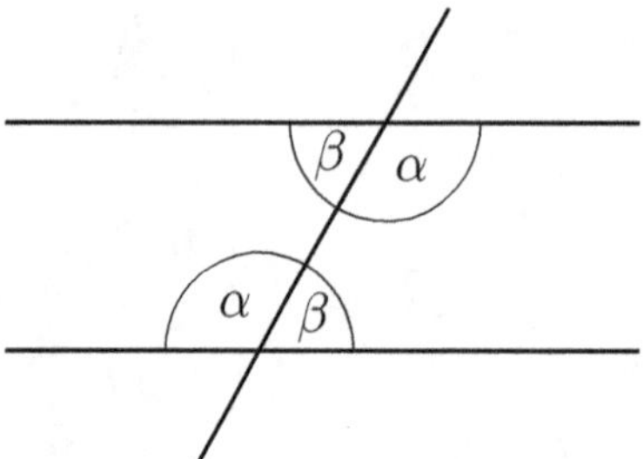

5.2   Lines

The most general equation, as you know, of a straight line is

$$ax + by + c = 0$$

If $a = 0$ the line is of course horizontal, and if $b = 0$ then the line is vertical.

The equation of the line can also be written in what is called the intercept form is

$$y = mx + b$$

where m is the slope or gradient. Setting x to zero in this equation will obviously give us the value at which the line passes the y-axis. Which is why it is called the intercept form.

An equation of the form $y = b$ wold be a horizontal line, and the equation $x = a$ is of a vertical line (which of course does not intercept y).

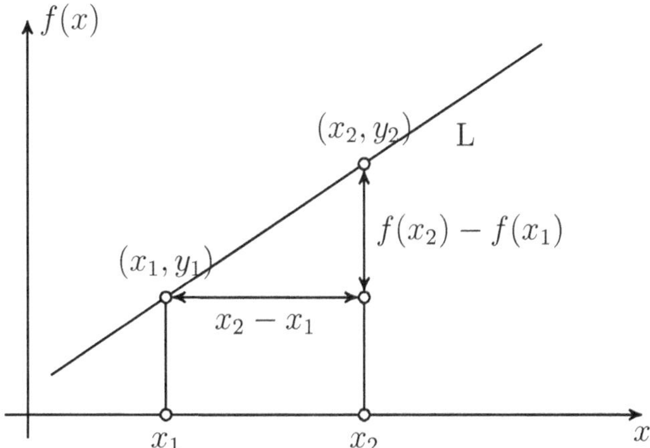

To calculate the slope we take two points on the line (x_1, y_2) and (x_2, y_2) and compare the vertical difference compared to the horizontal difference or distance.

You know that to find the length of a line segment you subtract the value of point one from the value of the second point, so that, for example, the length of a segment from $x_1 = 3$ to $x_2 = 8$ is $8 = 3 = 5$.

As in the graph above, a line passing between two points $p_1 = (x_1, y_1)$ and $p_2 = (x_2, y_2)$ will have a slope (some books call it gradient), and some write it as $\Delta y / \Delta x$

$$m = \frac{y_2 - y_1}{x_2 - x_1}$$

which you may remember as "rise over run."

It is not important which point is chosen as p_1 and which as p_2, but be careful that the order of the ys and of the xs be the same otherwise you'll get the wrong sign, negative instead of positive and vice versa.

Given two points (x_1, y_1) and (x_2, y_2) we can call their midpoint as in the figure (x, y):

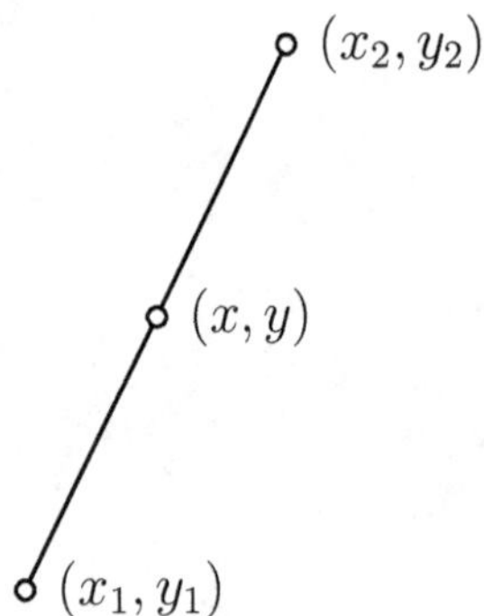

Since the slope of the line segment between the end points must be the same as the gradient of the segment from the midpoint to the end, we can set the equations for the two slopes as

$$\frac{y - y_1}{x - x_1} = \frac{y_2 - y_1}{x_2 - x_1}$$

and manipulating terms so that the ones with x are on one side we get the equation of a line passing through two given points:

$$\frac{y - y_1}{y_2 - y_1} = \frac{x - x_1}{x_2 - x_1}$$

And the distance between the points (x_1, y_1) and (x_2, y_2) is

$$d = \sqrt{(x_2 - x_1)^2 + (y_2 - y_1)^2}$$

which obviously follows from the Pythagorean theorem

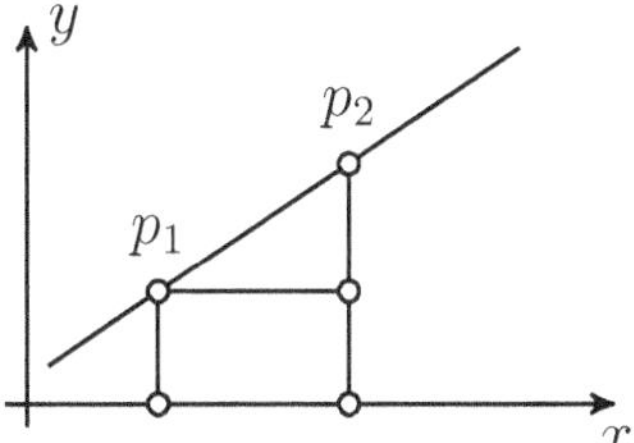

and their midpoint is

$$m = (\frac{x_1 + x_2}{2}, \frac{y_1 + 2_2}{2})$$

If we have two linear equations the lines may cross, as in the system:

$$y = 2 \cdot x + 3\frac{2}{3} \text{ and } y = \frac{1}{2} \cdot x + 4$$

which are the lines shown in the graph,

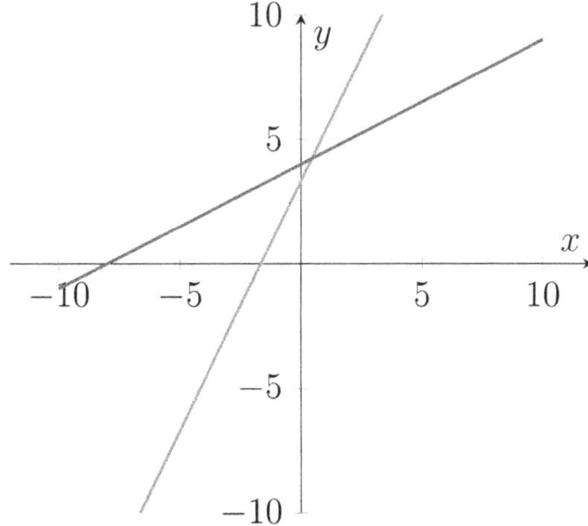

The point of intersection is the solution to the system of equations. Since the same x will yield the same value of y in both equations.

Rewriting the equations in the form $m \cdot x + n \cdot y = p$

$$y - 2 \cdot x = \frac{10}{3} \quad \text{and} \quad y - \frac{1}{2} \cdot x = 4$$

We can then solve the system by multiplying the first equation by -1, obtaining the two equations

$$-y + 2 \cdot x = -\frac{10}{3} \quad \text{and} \quad y - \frac{1}{2} \cdot x = 4$$

and adding the equations, so that the ys cancel out, letting us solve for x,

$$\frac{3}{2}x = \frac{2}{3} \rightarrow x = 1$$

Two linear equations, on the other hand, have no solutions if they are parallel, that is, they have the same slope and different y-intercepts.

$$\frac{1}{2}y = x = 1 \quad \text{and} \quad y - 2x = 5$$

As you can see by putting them in the intercept for.

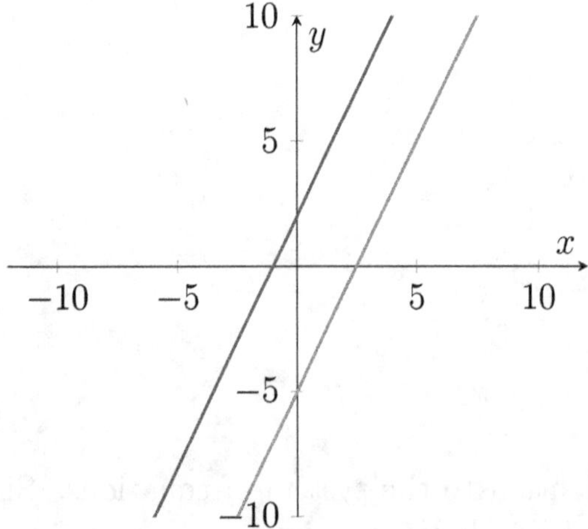

In case the lines have the same slope and the same y-intercept they are then, of course, the same line (in other words one of the two equations is simply the multiple of the other). In this case there are infinite solution since every (x, y) will correspond to the same (x, y) of the other line.

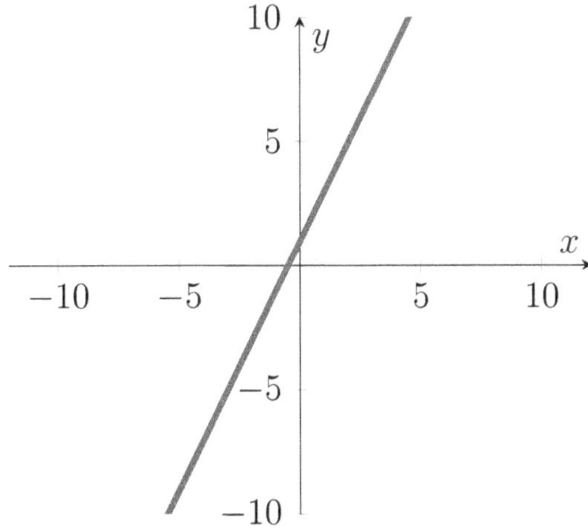

5.3   Exercises

Answers are in the back of the book, Appendix C.

1. What is x in the solution to the system of equations

$$3x = 21 - 5y$$
$$2x - 10y = 14$$

2. What is the slope of a line perpendicular to the line with equation

$$5y - 3x = 45$$

6

Quadratic Equations

6.1   Parabolas

Quadratic equations are of the form $f(x) = ax^2 + bx + c$

There are several approaches to solving a quadratic equation. You'll probably have learned the quadratic formula, which we'll get to in a shortly.

A quadratic equation $y = ax^2 + bx + c$ is a parabola, facing up

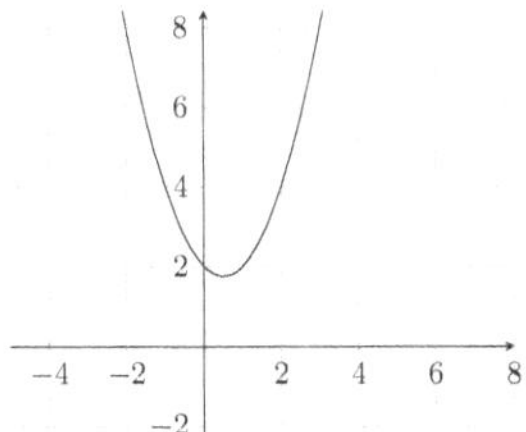

if the coefficient of x, that is a, is positive, think :-) and facing down if the coefficient of x^2 is negative.

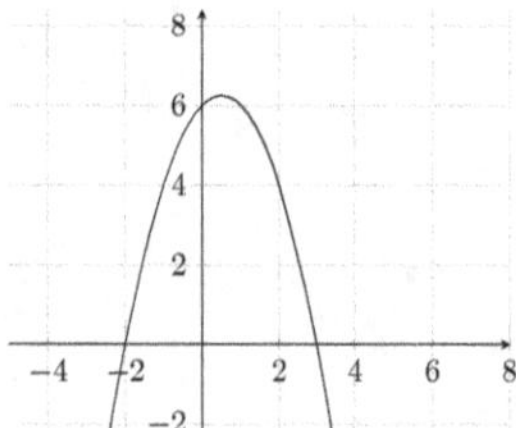

In many of the test problems the factors (or divisors) of the polynomial can actually be found by inspection.

Looking at the polynomial $x^2+B\cdot x+C$, if the coefficients B is the (algebraic) sum of two numbers n and m and and C is the product of the same two numbers $n \times m$, the two numbers will be the constants in the factors of the polynomial.

For example:

$$x^2 + 5x + 6 = 0$$

The factors of 6 are $\{6, 1, 3, 2\}$, and of these 3 ,2 add up to 5, so the divisors are $(x + 3)$ and $(x + 2)$. Make sure you observe the signs of the coefficients, however.

If the equation had been $x^2 - 5x + 6 = 0$, then the divisors of 6 could both be negative, and the divisors would then be $(x-3)$ and $(x-2)$. And of course if the constant is negative then its factors would be opposite signs and the coefficient of x would be their difference.

For example $x^2 + x - 6 = 0$, and in this case the divisors of the quadratic are $(x + 3)$ and $(x - 2)$.

When finding the roots of a polynomial see if you can factor out a factor common to all the terms first.

$$5x^2 + 15x + 25 = 0 \rightarrow 5(x^2 + 3x + 5) = 0$$

If the constant term and the coefficient of the middle term are both positive necessarily both factors of the positive term will be positive. The first example:

$$x^2 + 9x + 14 \rightarrow 9 = 7 + 2, 14 = 7 \cdot 2 \rightarrow (x + 2)(x + 7)$$

and here is the second example:

$$x^2 + 2x - 15 \rightarrow 2 = 5 - 3, 15 = 5 \cdot 3 \rightarrow (x + 5)(x - 3)$$

Rarely you will have a polynomial in which the leading coefficient is not 1 in which case a possible method for factoring trinomials of the type $Ax^2 + Bx + C$

Multiply the leading coefficient A and the constant C. Try to factor the product AC so that the sum of the factors is B. That is, find integers p and q such that $pq = AC$ and $p + q = B$. Factor by grouping, that is finding the common binomial.

$$5x^2 + 13x + 6 = 0$$

$5 \times 6 = 30$, we need to find factors of 30 that add up to 13, which can easily be seen to be 10 and 3., so we write $5x^2 + 3x + 10x + 6$, factor out the common binomial in $5x^2 + 10x$ and in $3x + 6$ we have $5x(x + 2)$ and $3(x + 2)$ which we can then write as

$$(5x + 3)(x + 2) = 0$$

You may want to practice with $15x^2 + 22x + 8 = 0$

In even fewer cases you if you need to use **the quadratic formula**, here it is (but which you probably remember):

$$x = \frac{-b \pm \sqrt{b^2 - 4ac}}{2a} \tag{6.1}$$

In the examples above $a = 1$. Using the quadratic formula then, the roots of the first polynomial in the equation $x^2 + 9x + 14 = 0$ are

$$x^2 + 9x + 14 \rightarrow x = \frac{-9 \pm \sqrt{9^2 - 4 \cdot 14}}{2} \rightarrow -2, -7$$

$$x^2 + 2x - 15 \rightarrow x = \frac{-2 \pm \sqrt{2^2 - 4 \cdot (-15)}}{2} \rightarrow -5, 3$$

Let us look at another way of considering a polynomial, with its graph:

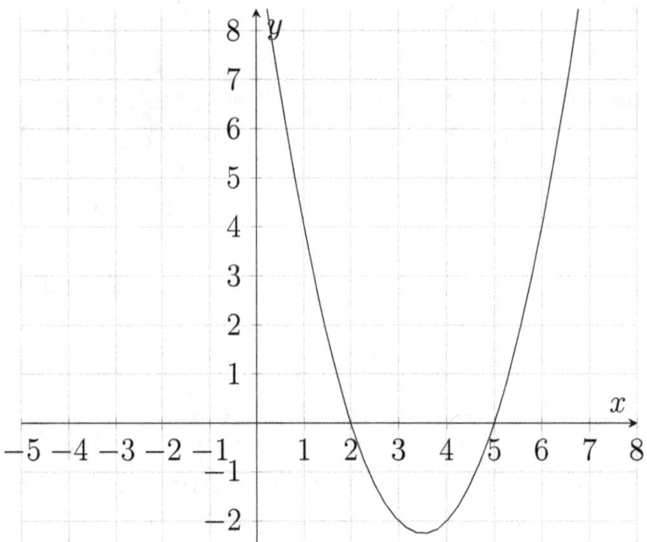

This is a second degree polynomial, and it has two real solutions (it crosses the X axis twice). In fact, by inspection, it looks like the solutions are 2 and 5, so the polynomial would be:

$$(x - 2)(x - 5) = x^2 - 7 \cdot x + 10 = 0$$

Aside from the general formula:

$$y = ax^2 + bx + c$$

a quadratic may also be expressed by the vertex formula.

$$f(x) = a(x - h)^2 + k$$

where (h, k) are the coordinates of the vertex, and in particular h is the x coordinate of the vertex which can be found with the simple formula

$$x_v = \frac{-b}{2a}$$

By putting that in the general equation $y = ax^2 + bx + c$, we have,

$$y = a(\frac{b^2}{4a^2}) + b(\frac{-b}{2a}) + c = \frac{4ac - b^2}{4a}$$

There is no need to memorize this last equation, since you can usually just put the value of $h = (-b)/(2a)$ into the equation you were given originally and obtain the desired y.

It is instructive to derive the vertex formula from the standard form and vice versa at least once.

$$y = ax^2 + bx + c$$

$$y = a\left(x^2 + \frac{b}{a}x\right) + c =$$

completing the ssquare $=$

$$y = a\left(x^2 + \frac{b}{a}x + (\frac{b}{2a})^2\right) + c - \frac{b^2}{4a} =$$

$$y = a(x + \frac{b}{2a})^2 + \frac{4ac - br}{4a}$$

The coordinates of the point at the vertex are then

$$\left(\frac{-b}{2a}, -\frac{b^2 - 4ac}{4a}\right)$$

And given the vertex formula, we can derive the standard form.

$$a\left(x - h\right)^2 + k =$$
$$ax^2 - (2ah)x + (ah^2 + k)$$

giving us $b = 2ha$ and $c = ah^2 + k$

As an exercise try proving

$$a(x - 1)^2 + 33 = 2x^2 + 4x + 5$$

The x coordinates may also be found as the average of the "zeroes".

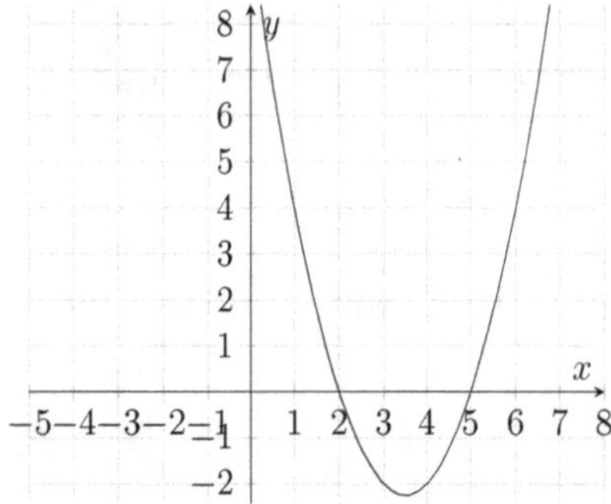

In the parabola we see that the x coordinate of the vertex is exactly between 5 and 2, that is, at 3.5, so for this parabola, whose equation is $y = x^2 - 7x + 10$, we substitute 3.5 or $7/2$, and we have

$$y = \left(\frac{7}{2}\right)^2 - 7\left(\frac{7}{2}\right) + 10 = -\frac{9}{4}$$

as you can verify from the graph. Actually, you could use any horizontal line segment from one side of the parabola to the other, for example, again looking at the graph, the segment at $y = 4$, which is from $x = 1$ to $x = 6$, the midpoint, again, is at $x - 3.5$

For higher powers note that polynomials with odd powers are asymmetrical, that is $f(x) = f(-x)$, and whose range (y values) can go from minus infinity to positive infinity, as in the example $y = x^3 - 9x$:

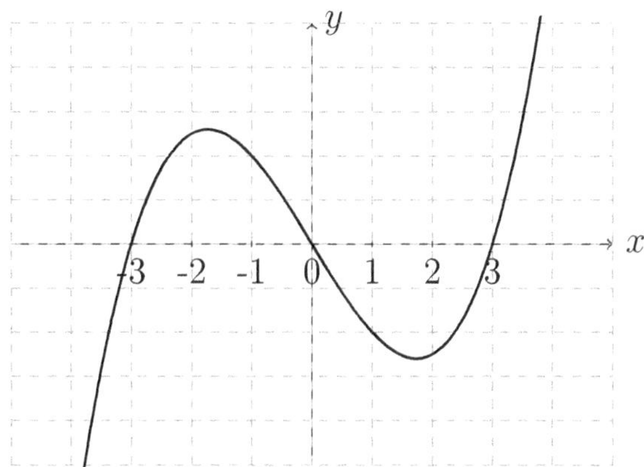

and with only even powers the graph may be symmetrical, i.e. $f(x) = f(x)$, but in any case it will have a maximum or minimum y value. Here is the graph

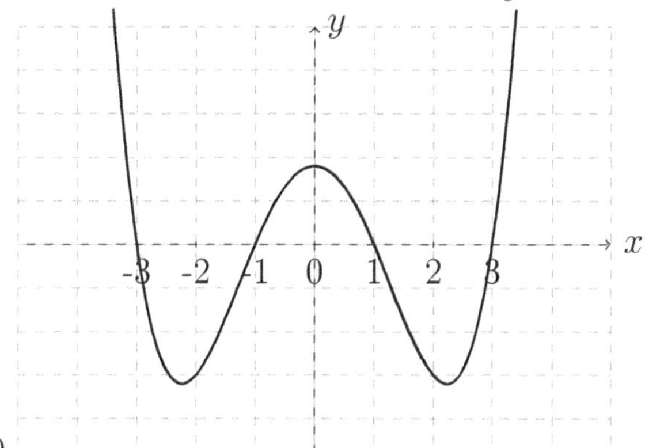

of $y = x^4 - 10x^2 + 9$

There will only be a few problem with polynomials of degree higher than two. You will not be expected to know any formulas. The approach will be to factor out some power of x (and a coefficient) and then manipulate the remaining polynomial of second degree.

For example:

$$x^3 - 9x = x(x^2 - 9) = x(x + 3)(x - 3)$$

System of equations

We saw with linear equations that the point x, y where the lines cross is the solution to the system of equations. I hope it's clear why. Given that value of x in both equations will give the same y, so that point is their common solution.

This concept is the same for any degree of polynomials. So the points where a line crosses a parabola will be the solution to the system of two equations.

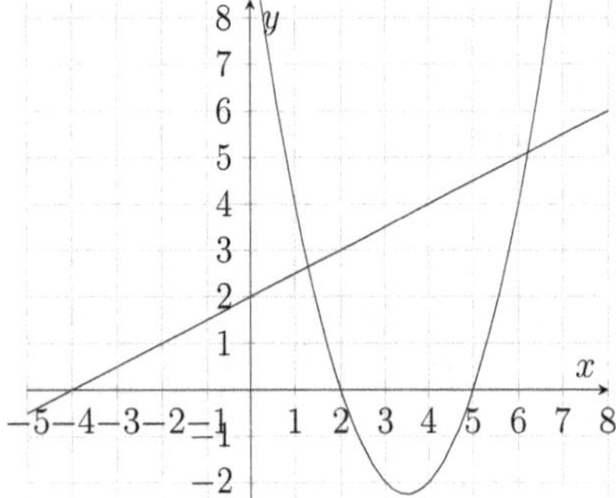

The two equations plotted in the figure cross at the two points and thus the system has two solutions. Whereas in the figure below the line is tangent to the parabola and the system has only one solution.

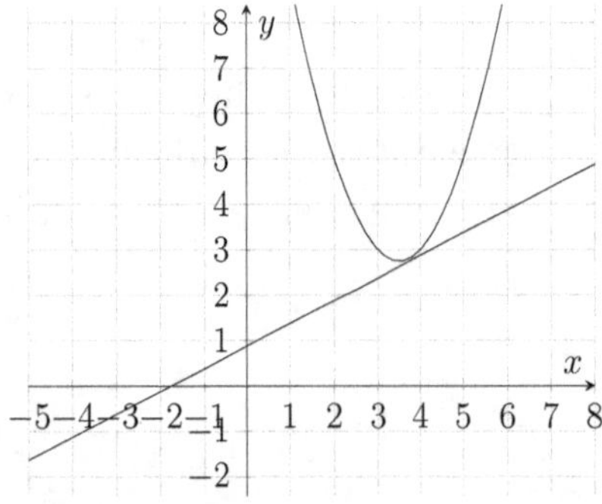

If the linear equation and the quadratic have no points in common then the system has no solution and the graphs of the two equations do not meet.

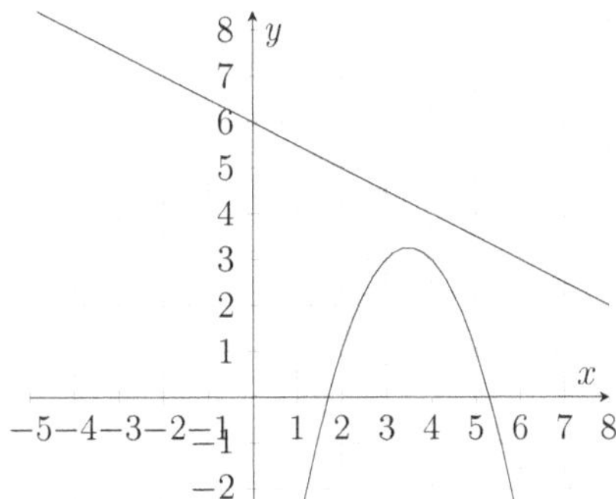

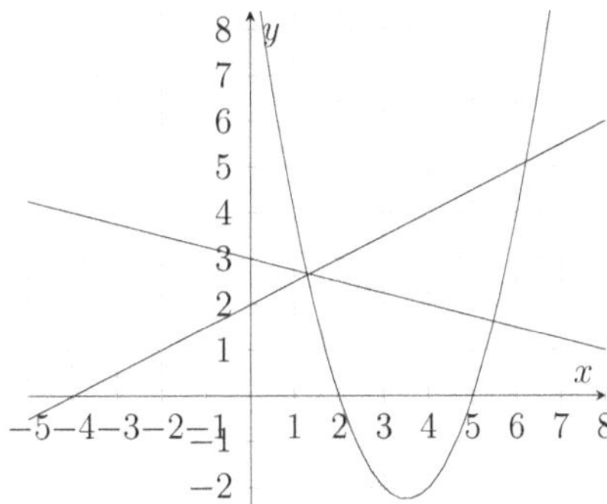 You may be shown the plot of more than two equations. The system of equations has a solution only if there is a point common to *all* the plots.

In the graph below there is one, and only one, solution to the system of three equations, one quadratic and two linear. Can you see where the point representing the solution lies?

6.2 The Circle

If a circle is centered in the origin, that is, its center point is at $(0,0)$ and its radius is r, the circumference will always be distance r from the center.

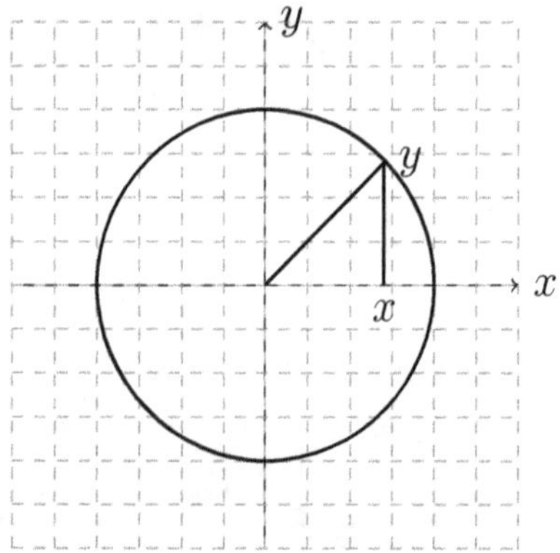

If we pick any point (x, y) at all on the circumference, the segment of length x from the origin, and the segment of length y, forming a right angle, satisfy the Pythagorean theorem with the resulting hypotenuse, from $(0, 0)$ to (x, y).

The equation of this circle then is exactly $x^2 + y^2 = r^2$. Since this is true for any point on the circumference this equation then defines this circle.

Now let us imagine that we shift the circle k units to the right,

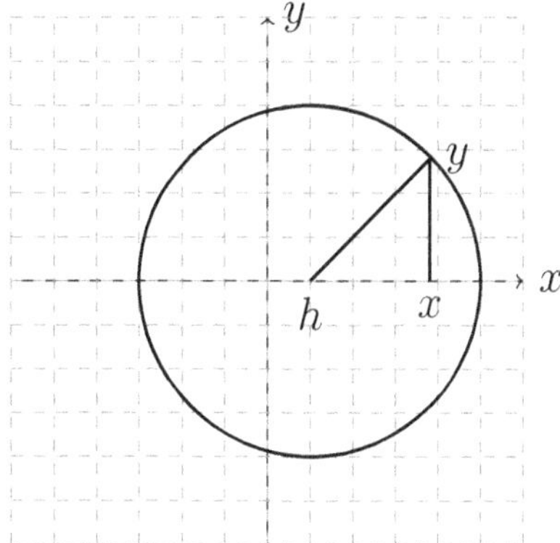

The center of the circle is now at h along the x-axis, and the length of the lower leg of the triangle would be $x = h$. So we must subtract the extra segment of length h, and our equation for the circumference becomes $(x - h)^2 + y^2 = r^2$

Shifting the circle once more, say down k units, we find that to obtain the length of the vertical leg of the triangle we must add k units.

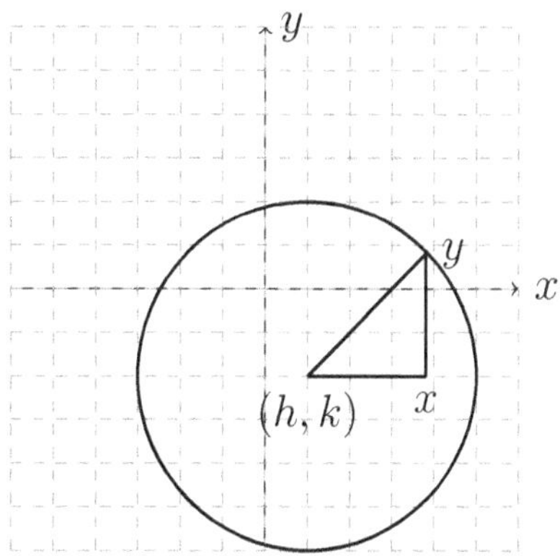

The general formula therefore is:

$$(x - h)^2 + (y - k)^2 = r^2$$

Wait! You might ask, are the signs right, in the example we added k.

Because it is an algebraic subtraction, $y - (-k) = y + k$

6.3 Exercises

Answers are in the back of the book, Appendix C.

1. Simplify
$$8\frac{720d^2}{9d^2(-8d)} = x$$

2. Simplify
$$x^2 + 6x + 7 = 1 - x - x^2$$

3. Simplify
$$x\left(\frac{1}{2}\right)(2x+3)^{(} - \frac{1}{2})(2) + \sqrt{2x+3}$$

4. Solve the inequality:
$$-2 < 3x + 1 < 3$$

5. Solve the inequality:
$$2x - 1 \le 3x - 5 \le x + 9$$

6. Find the factors
$$10x^2 + x - 3 = 0$$

7. Simplify

$$\left(\sqrt{4x} + \sqrt{9x}\right)^2$$

7

Growth, Decay

7.1   Exponential Functions

Legend has it that the sheikh or king was so impressed with the invention of the game of chess that he told the inventor to name his reward. The inventor replied,

"I just want one grain of rice on the first square, two on the second, four on the third, eight on the fourth, and so on ..."

The king was amazed. He could not believe the person's humility, until a few days later the scribes informed the king that not only was there not enough rice to pay the inventor, but even after many centuries it would not be possible to grow that much. (No one has reported how the king reacted, but being outsmarted by a mere subject, it must not have been very well.)

It was probably not the first time, and certainly not the last that people are fooled by exponential functions. If we look at the graph of two simple functions, one linear and the other exponential we can see that the exponential function seems to grow slower than the linear at first.

Maybe that is why in many countries politicians were fooled when the corona virus began, since it did not seem like cases were increasing until too many people had been infected and hospitals were overwhelmed, and only then did they realize the gravity of the situation.

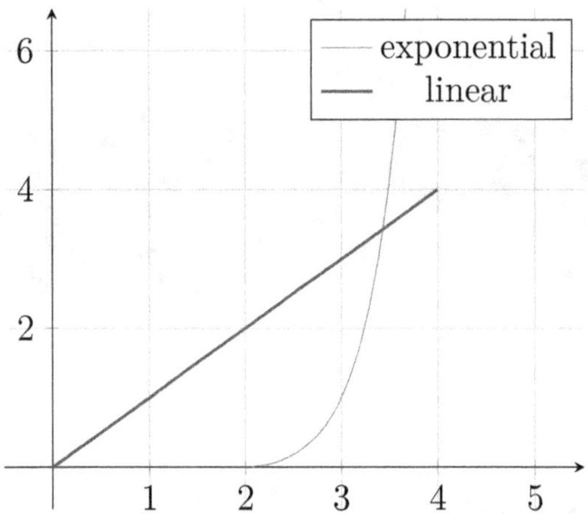

The most common type of questions involving exponential functions in the tests have to do with growth, compound interest or decay.

If the problem states that a bank gives a person a percentage of the initial amount every year, that is not an exponential growth, but rather a linear growth.

For example if John deposits 500 $ and the bank adds 3% of the initial investment (that is of the 500 $) then the growth will be 500, 515, 530, 545 and so on, an the formula would be

$$y = 15 \cdot x + 500$$

where y is the final amount after x years, and 15 is the fixed rate ($500 \cdot 0.03 = 15$). That is simple interest.

Compound interest means that it is proportional not to the initial amount but to the latest amount, thus the base amount and the amount added increases every year.

As usual, rather than diving into the formula let's see how it follows naturally and logically.

Let us take a typical problem. Jane invests 3200 dollars in a bank at 3% interest, how much will she have after 6 years? Don't worry about the numerical value, but see if you can write the solution as an expression.

At time $t = 0$ the amount will be 500 dollars. At the end of the first year Jane will have:

$$y = 500 + 500 \cdot 0.03$$
$$= 500(1 + 0.03)$$
$$= 500(1.03)$$

I hope you noticed how we used the distributive property, because we will continue using it in this process:

$$1 \cdot 500 + 500 \cdot 0.03 = 500(1 + 0.03)$$

The second year we begin with the new amount $(500 \cdot (1.03))$:

$$y = 500(1.03) + 500(1.03) \cdot 0.03$$
$$= 500(1.03)(1 + 0.03)$$
$$= 500(1.03)(1.03)$$

and the third year

$$y = 500(1.03)(1.03) + 500(1.03)(1.03) \cdot 0.03$$
$$= 500(1.03)(1.03)(1 + 0.03)$$
$$= 500(1.03)(1.03)(1.03)$$

proceeding in this fashion, the following years will be $500(1.03)(1.03)(1.03)(1.03)$ and so on. Which we of course can write as $y = 500(1.03)^t$, where t is the number of years. In the problem as stated above the answer would be $500(1.03)^6$

Below is the graph for $500(1.03)^t$ where t is from 0 to 10 years. Note that for such a short period of time the curve is not easy to see.

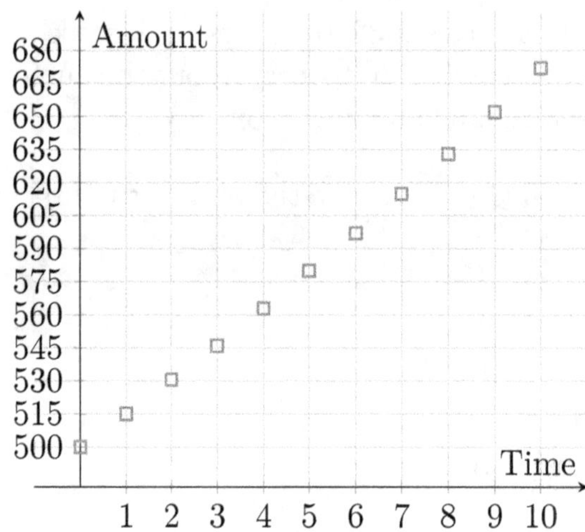

When given a graph, make sure to pay careful attention to the units used in the x and y axis. The y axis may be labeled $1, 2, 3 \ldots$ but the units may be thousands.

Or it may not even be a linear scale. The same information on a logarithmic scale for y is a straight line, when the same distances on the y along the axis are actually for increasing values as you go up the scale.

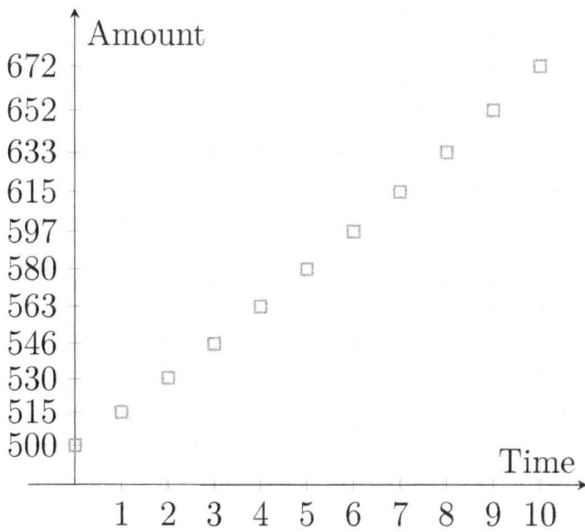

7.2   Exercises

Answers are in the back of the book, Appendix C.

1. If someone deposits 2,200 $ in a bank that gives 2.3 % interest what is the equation for the amount after m years?

2. How much will you have after 3 years if you deposit 3,000 dollars in a bank that gives you 5% interest? (Round to the nearest cent.)

8

Geometry

8.1   Plane Figures

You will find that it is indispensable to know your triangles and circles, as well as regular plane figures, and quadrilaterals if you want to get a decent score.

The Pythagorean theorem is the one that underlies many of the geometric problems: the square of the hypotenuse (the side opposite the right angle) is equal to the sum of the squares of the other two sides, and it sometimes helps to calculate some special relationships when you cannot recall them.

Triangles

For the SAT there are really four triangles you will need to know well, as we'll see below, they are what we could call the 45-45-90, the 30-60-90 triangles (referring to their internal angles) and the 3-4-5, and the 5-12-13 triangles (referring to the ratio of the sides).

Be on the lookout for them when you are given a problem involving triangles.

It will make life easier.

✎ **Remember these ratios: 3-4-5 and 5-12-13, and these angles, 30-60-90 and 45-45-90**.

The Triangle rule: the length of a side must be between the difference and the sum of the other two.

Obviously, if the sum of two sides is less than the remaining side they could not possibly form a triangle.

As we've seen over and over, the Pythagorean theorem is one that underlies many geometric problems: the square of the hypotenuse (the side opposite the right angle) is equal to the sum of the squares of the other two sides.

It is a good idea to calculate some special relationships on your own so that you learn to quickly recognize them.

The Pythagorean Theorem

$$a^2 = b^2 + c^2$$

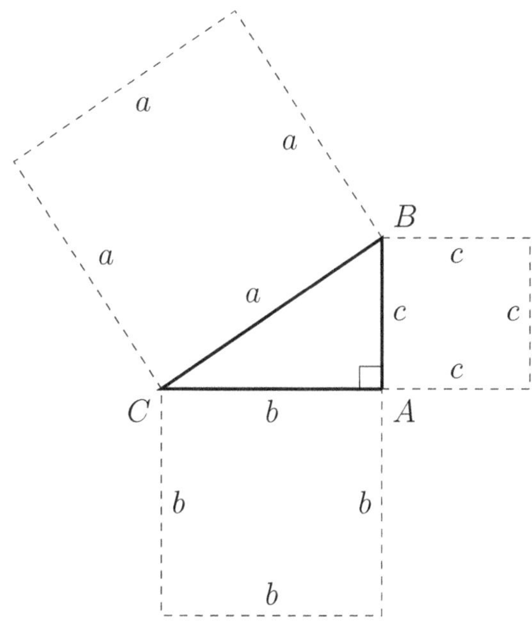

The formulas for areas and perimeters of plane figures, and area and circumference of a circle should be memorized. There aren't many you need to know for the test.

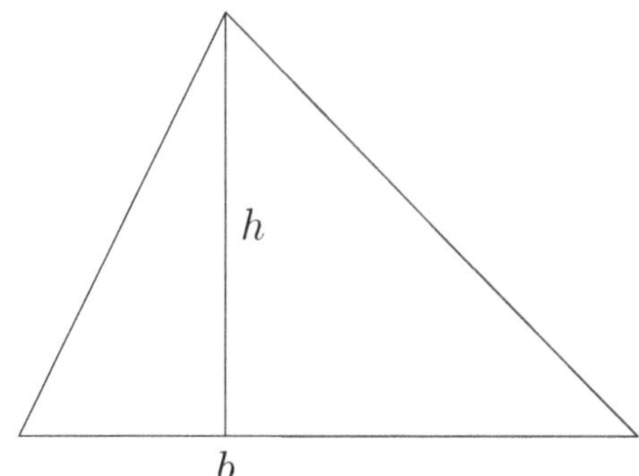

The area of a triangle is one half the base times the height. One consequence is that two triangles drawn between two parallel lines, with the same base, whose apex lies on the other line will have the same area.

$$A = \frac{b \cdot h}{2}$$

Some problems will involve looking at the ratio of the sides of a triangle, commonly the ratios $3 : 4 : 5$, and sometimes $5 : 12 : 13$, these are always right triangles.

If the base of a wooden triangle were 12 m long, its perpendicular height 9 m, and the remaining side 15 m then it must be a right triangle.

So when you are given two sides of a right triangle, or given three sides, take a moment to see if the respect the above ratios so that the problem may be solved without calculating the sides.

Two triangles which are often included in problems are not with the ratios of their sides but with the value of the angles. These two are (in degrees) the 45-45-90 and 30-60-90 right triangles.

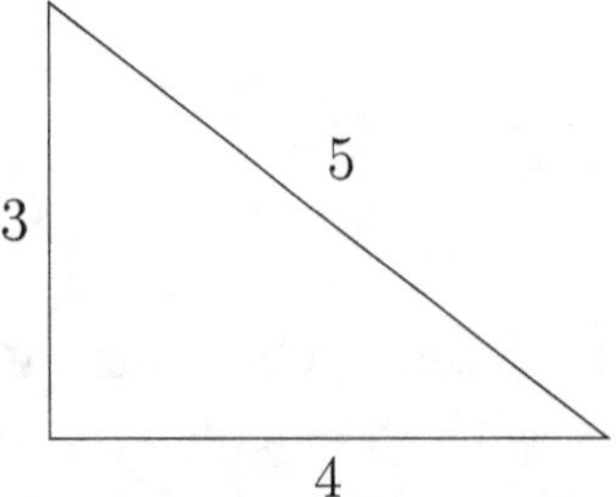

The 45-45-90 degree triangle is an isosceles triangle,

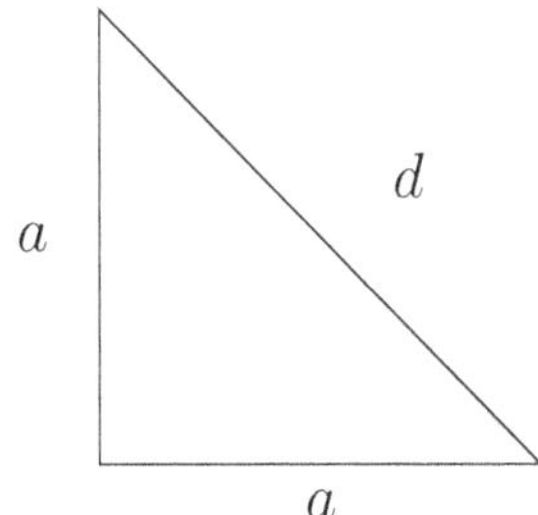

and the 30-60-90 degree triangle, with sides of differing lengths.

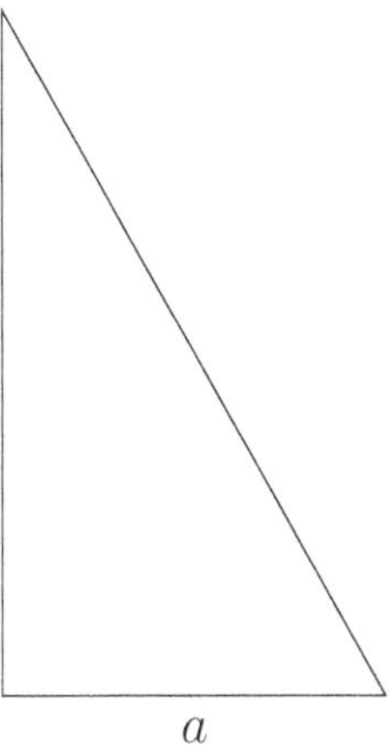

Let us take a look at the ratio of the sides of these triangles, which, once you've learned them by heart will save a lot of calculations.

Note that the first triangle, 45-45-90, is one half of a square (the hypotenuse being the diagonal).

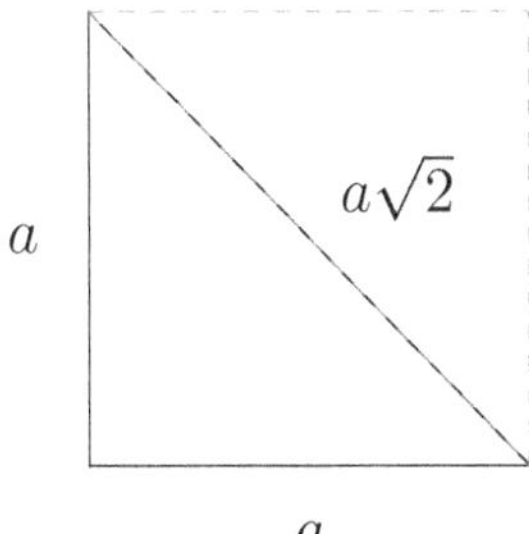

And the diagonal can be calculated given a as one of the sides: $a^2 + a^2 = d^2 \rightarrow 2a^2 = d^2 \rightarrow \sqrt{2a^2} = \sqrt{2} \cdot \sqrt{a^2} = a \cdot \sqrt{2} = d$

What would the hypotenuse be of a 45-45-90 triangle whose legs are 5 cm long? (Remember the hypotenuse of such a triangle can be imagined to be the diagonal of a square.)

Of course, it's $5\sqrt{2}$. If you've learned what the sides of this triangle are in general, you can solve any of them without using the Pythagorean formula. Isn't that great?

What if you were given an isosceles right triangle whose hypotenuse is $18\sqrt{2}$, what is its area?

First you would see that such a triangle can only be a 45-45-90 triangle. (Can you verify that?)

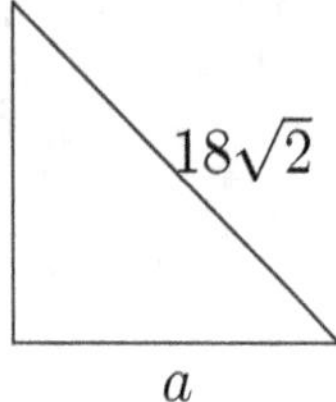

Secondly, you know that the ratio of a leg to the hypotenuse is $a : a\sqrt{2}$, so thus triangle it's $18 : 18\sqrt{2}$, which means that the length and height of this triangle are both 18, and using the formula for the area of a triangle: $\frac{bh}{2} = \frac{18 \cdot 18}{2} = 9 \cdot 18 = 162$

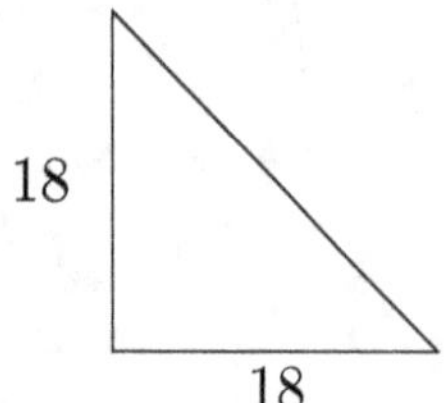

The second triangle. the 30-60-90 also crops up very frequently in tests.

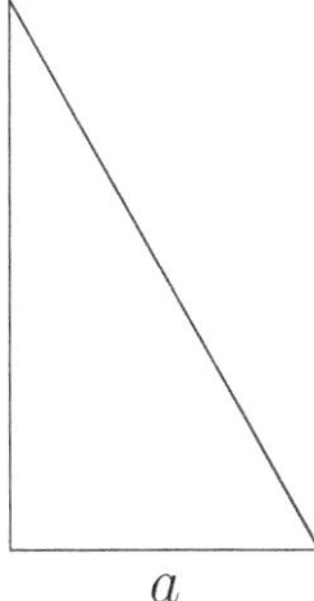

We can proceed similarly as we did for the first one. This time we duplicate the triangle along the side, to obtain a triangle with twice the area. Since the angles are all 60 degrees, the sides of the new triangle are all equal.

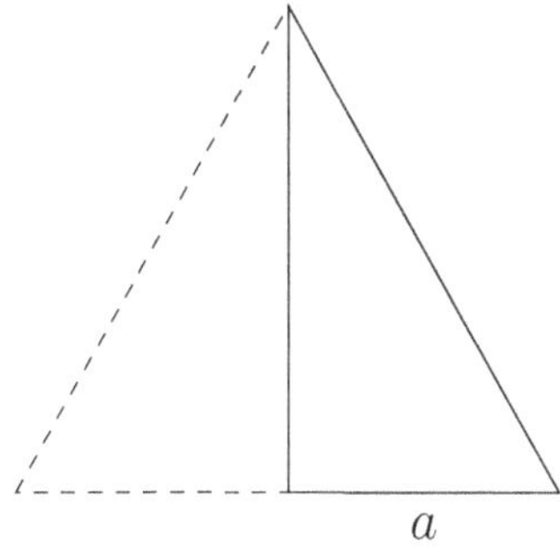

So obviously we now know the length of the hypotenuse:

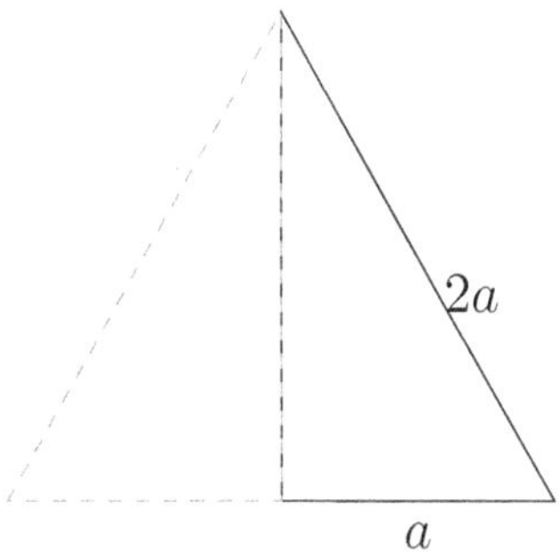

And proceeding as we did for the first triangle, but this time calculating the height, we have $h^2 + a^2 = (2a^2) = 4 \cdot a^2 \rightarrow h^2 = 4a^2 - a^2 = 3 \cdot a^2 \rightarrow h = a\sqrt{3}$

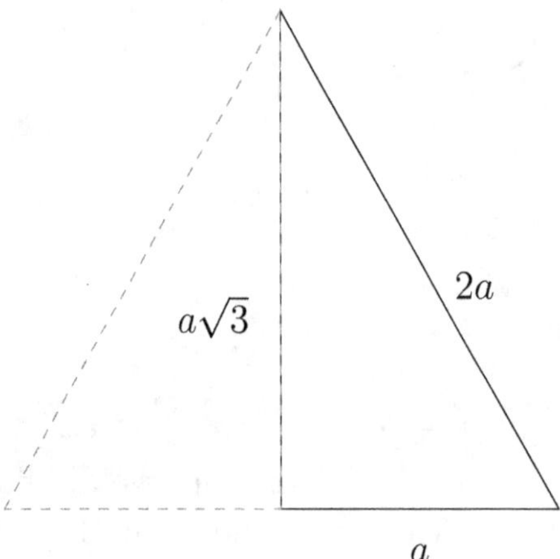

Let's see how we can apply what we know about these two triangles.

Suppose you are given the following triangle

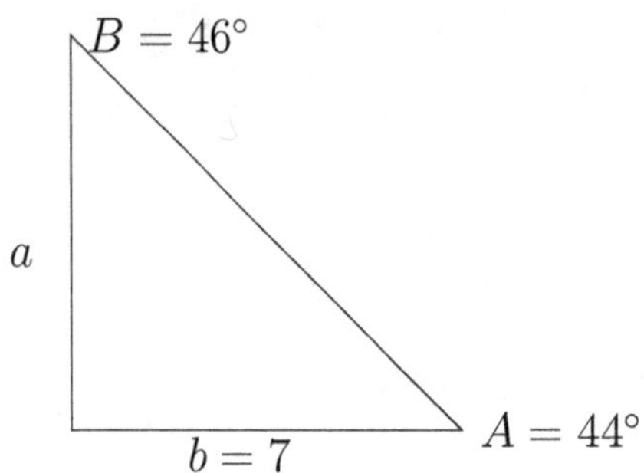

With side $b = 7$, and the opposite angle $B = 46°$, side a of unknown length, whose opposite angle is 44°, and you are asked whether the area of the triangle is less, equal to or greater than 24.5.

You could of course find the exact answer using the ratio of angles and sides:

$$\frac{a}{\sin(A)} = \frac{b}{sin(B)}$$

thus solving for a. This of course requires knowing the sin of angles A and

B, respectively 0.69466 and 0.71934 approximately, and giving the length of $a = 6.7598$.

From this we can calculate the area of the triangle to be about 23.6.

For obvious reasons, I hope, one can see this is not the best approach. The fastest way is to observe that the triangle is similar to a 45-45-90 triangle, the side of length 7 is opposite the 46° angle and the other leg is opposite a smaller angle, meaning that the side is less than 7. The right triangle whose (equal) legs of length 7 has an area of 24.5, thus the area of the given triangle must be less.

8.2   Polygons

The sum of the angles in a polygon of n sides is $S = (n - 2) \cdot 180$, and the sum of the external angles is 360. You can also find the sum if you don't remember the formula, draw diagonals from one vertex to obtain triangles.

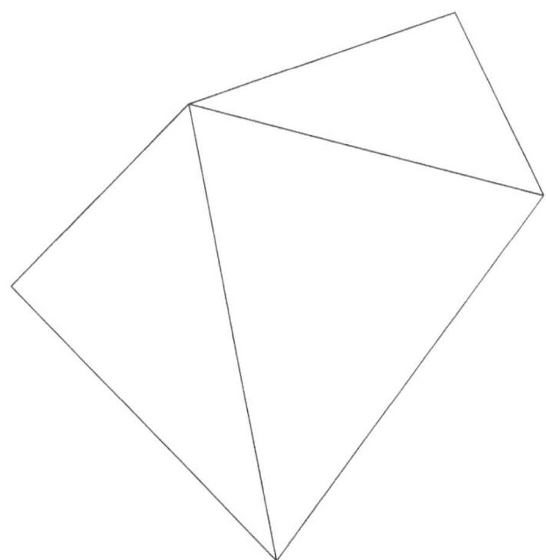

And the sum of the angles will equal 180° times the number of triangles formed. The formula for the sum of the internal angles, which you can derive graphically, where n is the number of sides, is:

$$\sum = (n - 2) \cdot 180$$

As the number of sides of a polygon increase as the sum of the internal angles also increases, the sum of the external angles cannot grow. And in fact, the sum of the external angles is, interestingly enough, always a constant $360°$

For a triangle this gives us an interesting property.

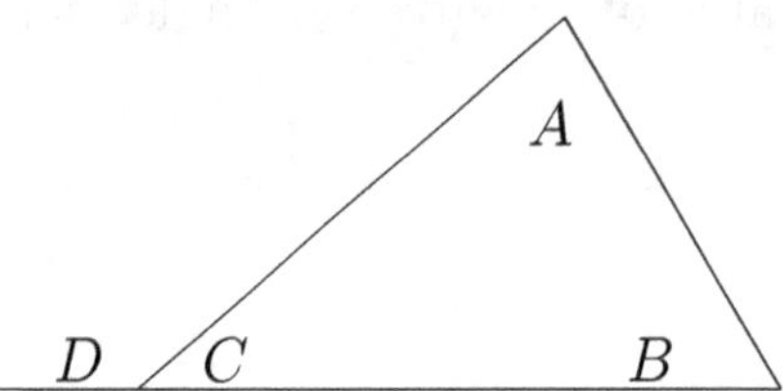

Since $A + B + C = 180$, and $D + C = 180$ we get that $D = A + B$, that is the external angle of a triangle is equal to the sum of the to opposite internal angles.

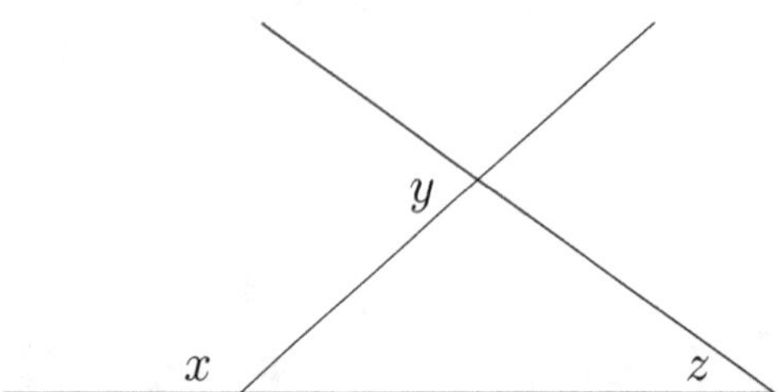

So given the figure above with angles $y = 110°$ and $z = 75°$, find angle x.

The supplementary angle to y is $70°$, so $x = (70 + 75)° = 145°$

Quadrilaterals

You are of course expected to know how to find the perimeter and area of a rectangle.

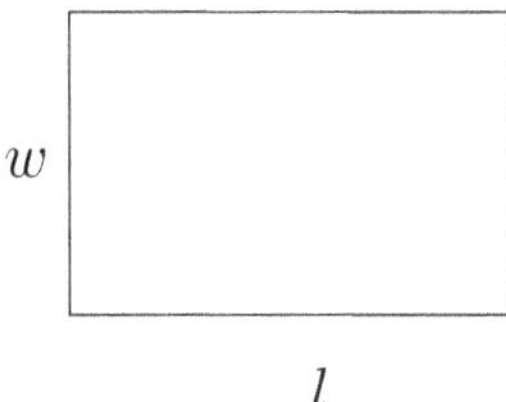

The area being width times the length, $A = w \cdot l$, and the perimeter $P = 2 \cdot w + 2 \cdot l = 2(w + l)$. And for a square all this of course simplifies, since w and l both equal s we have the area s^2, which is why a number times itself is called squaring, and the perimeter is simply $4 \cdot s$.

As we have seen, the diagonal of a square is $s\sqrt{2}$.

Another four sided figure that you may encounter is the rhombus, defines as an equilateral parallelogram. That is, its sides are equal and each is parallel to the opposite side.

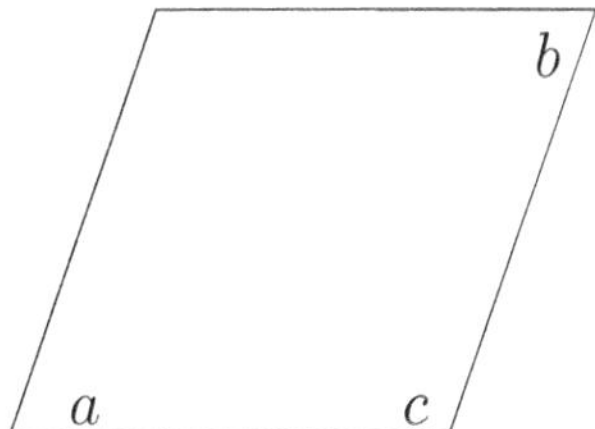

Note that the sum of the angles is 360, and $a + c = 180$ (That is adjacent angles are supplementary).

The square is a special rhombus whose angles all equal $90°$

The diagonals of a rhombus are perpendicular, they form $90°$ angles.

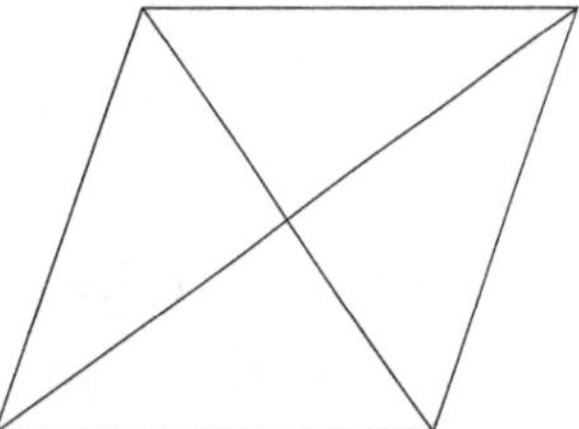

The trapezoid is a four sided figure having only two parallel sides. The other two are not necessarily equal in length.

The area of a trapezoid is the average of the two parallel sides times the height (that is the distance between the two parallel sides). Let's see how we can derive the formula.

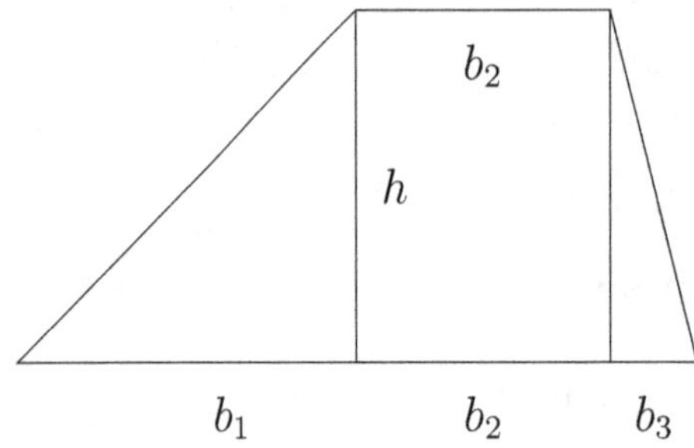

To see this let's divide the the trapezoid into three parts, the two triangles and the rectangle between them formed by the height drawn down from the top corners.The base of the triangle on the left, the rectangle and the triangle on the right we'll call b_1, b_2 and b_3 respectively. The top base is equal to b_2. So the area of the three figures will be

$$A = \frac{1}{2}b_1 \cdot h + b_2 \cdot h + \frac{1}{2}b_3 \cdot h$$

Now we will rewrite the middle term as $\frac{1}{2} \cdot 2 \cdot b_2$ so that we can factor out $\frac{1}{2} \cdot h$ from all the terms. We can then rewrite the equation as

$$A = \frac{1}{2} \cdot h \left(b_1 + 2 \cdot b_2 + b_3\right)$$

Since $b_1 + b_2 + b_3$ is the bottom base (let's call it B_1) and the top, b_2, we'll call B_2, we can rewrite the formula as

$$A = h \cdot \left(\frac{B_1 + B_2}{2}\right)$$

which is what we expressed in words.

8.3 Circles (again)

The irrational number π comes up in many equations. Recall comes from the ratio of the circumference to the diameter

$$\pi = \frac{C}{d} \rightarrow C = \pi \cdot d = 2\pi r$$

The reason we use the radius is that in a lot of formulas it is more natural than the diameter (which we would have to otherwise keep using the expression $\frac{d}{2}$, for example, for the area of the circle

$$A = \pi \left(\frac{d}{2}\right)^2 = \pi r^2$$

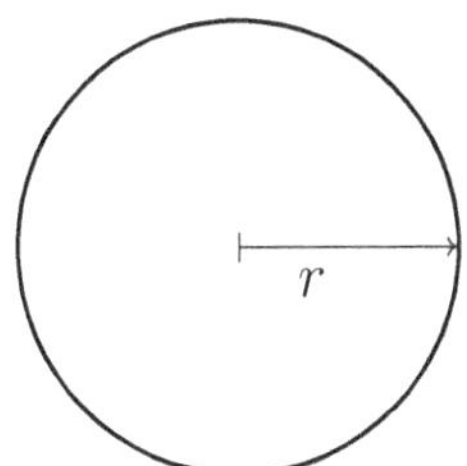

The arc of a circle is a portion of the circumference

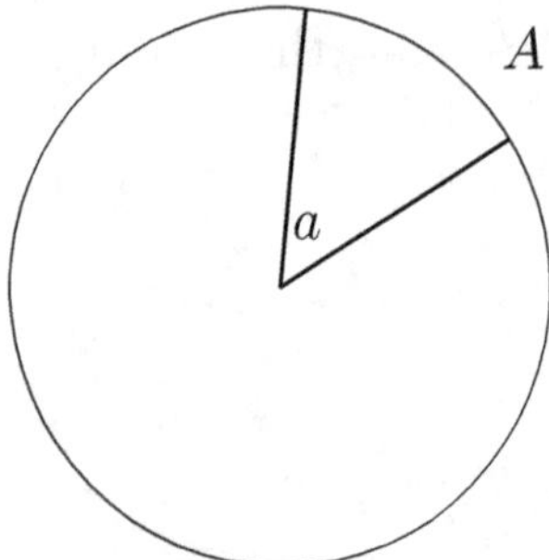

The ratio of the length of the arc, A, is equal to the ratio of the angle to $360°$, that is $A : C = a : 360$, so

$$A = \frac{a}{360} \cdot C = \frac{a}{360} \cdot 2\pi r$$

There are a number of important properties of circles that may prove to be useful in many problems. One of these is that any inscribed triangle whose base is the diameter of the circle, the angle tangent to the arc will be a right angle.

So in the picture below, the circle of diameter ab, both angles α and β are right angles.

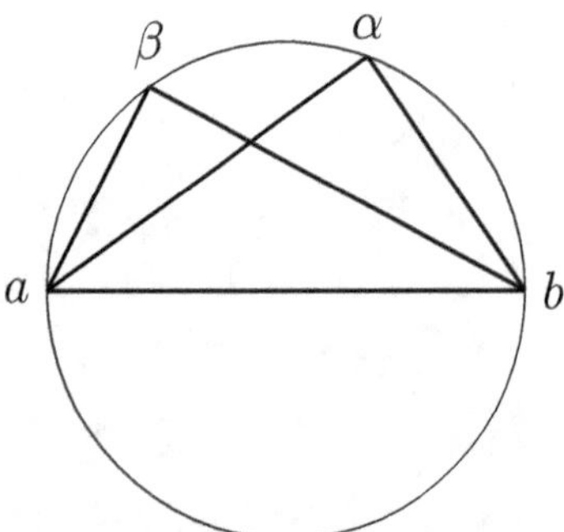

In the picture below,

$$x = \frac{1}{2}(A + B)$$

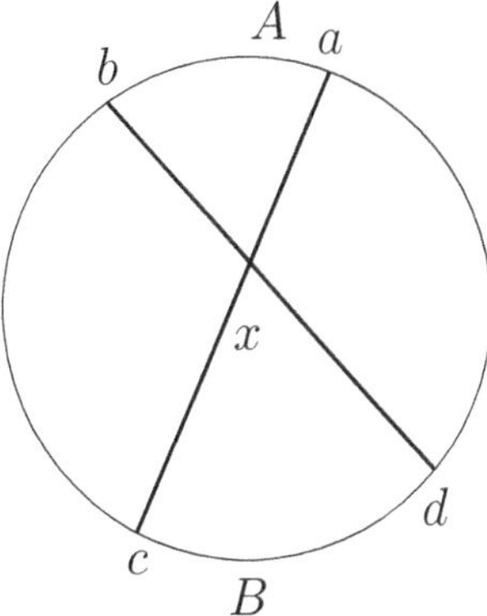

8.4 Solids

Starting with one of the simplest solids, a rectangular prism. The Volume is given by the area of the base times the height. The area of the base being $l \cdot w$, the formula then becomes simply $V = l \cdot w \cdot h$

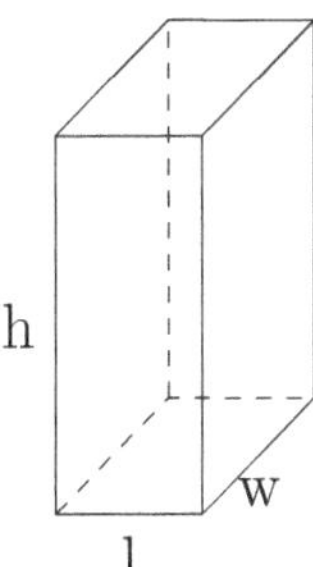

The volume of a cube then is simply $V = s^3$, where s is one of the edges of the cube.

Similarly, the volume of a cylinder is the area of the base times the height. The base being the circular shape of course, so with the area of a circle $\pi \cdot r^2$ we get the formula fora cylinder: $V = \pi \cdot r^2 \cdot h$.

The formula for the volume of a circular cone then becomes very easy, it is 1/3 the volume of a cylinder with the same base and height, that is $V = 1/3 \cdot \pi \cdot r^2 \cdot h$ And the same goes for a rectangular pyramid, it is 1/3 the volume of the enclosing prism.

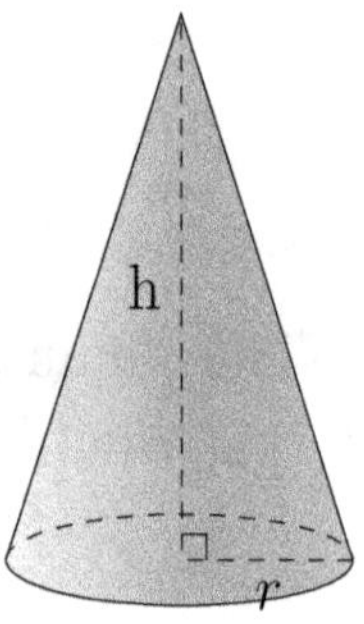

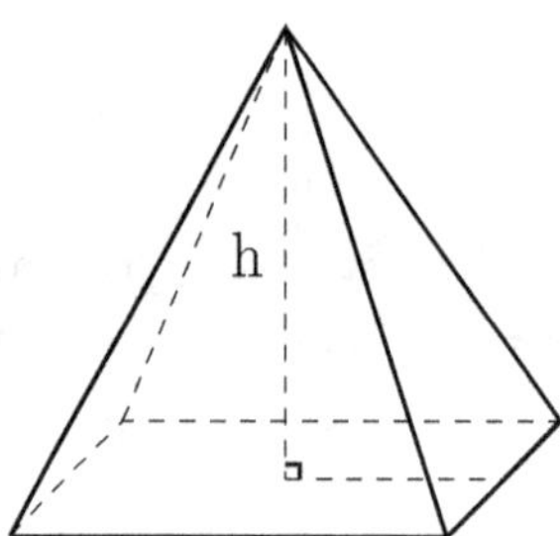

The volume of a sphere, just like the area of a circle, is proportional to the radius, however to its cube, and a factor o 4/3, so the formula should not be too difficult to remember, $V 4/3 \cdot \pi r^3$

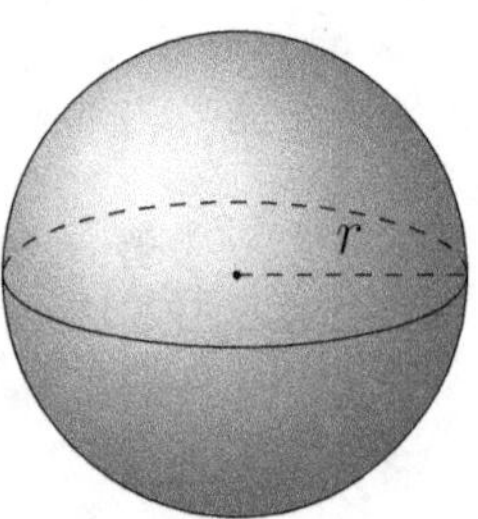

8.5 Exercises

Answers are in the back of the book, Appendix C.

1. A circle is inscribed in a square. A line from the center of the circle to a corner of the square is $3\sqrt{2}$ in length. What is the area of the area filled in gray?

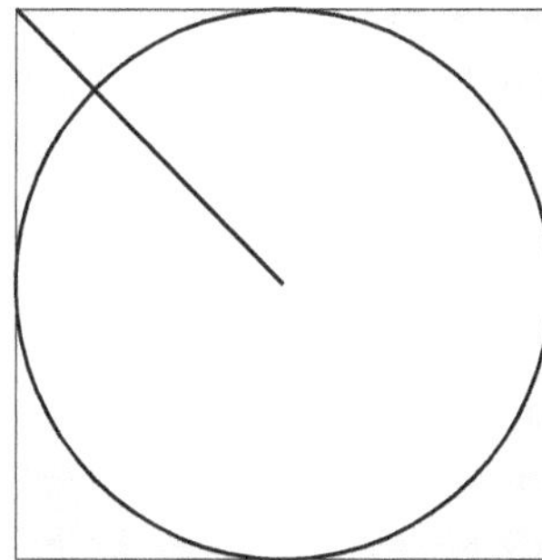

2. What is the area of a regular hexagon whose distance from the lower side to the upper side (line segment a) is $12\sqrt{3}$?

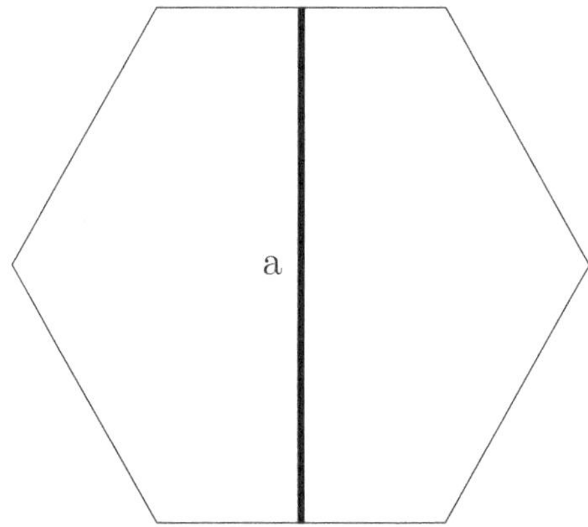

9

Trigonometry

9.1 Sin and Cos

In a unit circle $\sin\alpha = \frac{\text{rise}}{\text{radius}}$. It may help to remember that the "rise" tends to zero as it approaches the x-axis and $\sin(0) = 0$. Similarly $\cos\alpha = \frac{\text{run}}{\text{radius}}$ and of course the $\tan\alpha = \frac{\sin}{\cos} = \frac{\text{rise}}{\text{run}}$.

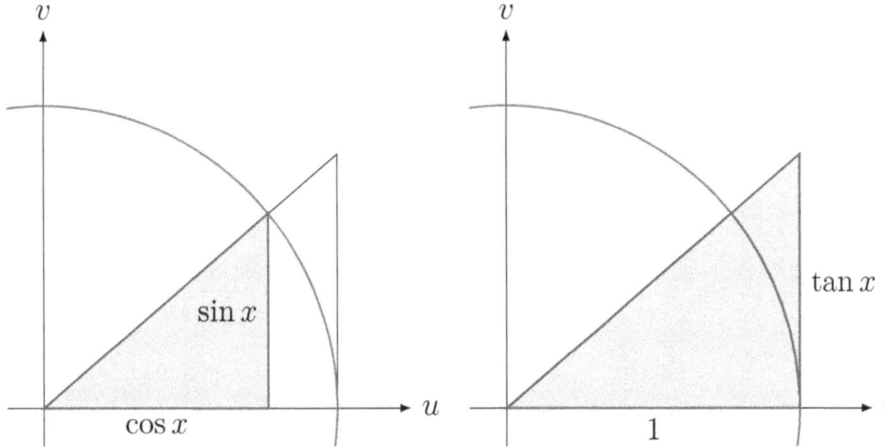

What are sin and cos? One answer was the ratio we just saw in the coordinate plane, another way of understanding them is as the graph of the y as the angle (x) increases.

Imagine the small dot circling counterclockwise, and for each angle x plot the height y of the dot

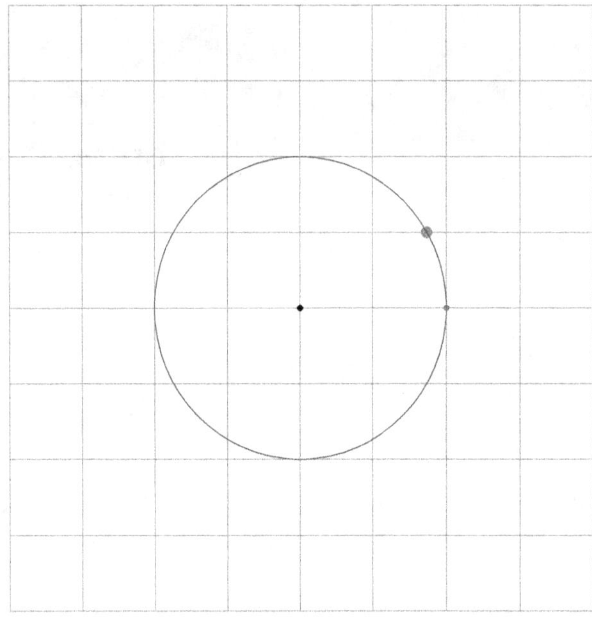

which will plot as the sine of the angle

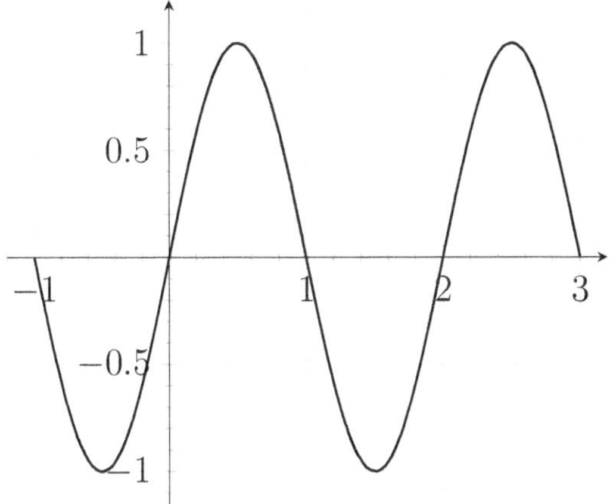

9.2   Radians

The magnitude in radians of one complete revolution (360 degrees) is the length of the entire circumference divided by the radius. So 2π radians is equal to $360°$, one radian is thus equal to $\frac{180}{\pi}$ degrees. Therefore, to convert from radians to degrees, multiply by $\frac{180}{\pi}$. Examples: $\frac{\pi}{6} \cdot \frac{180}{\pi} = 30°$ and $\frac{\pi}{4} \cdot \frac{180}{\pi} = 45°$

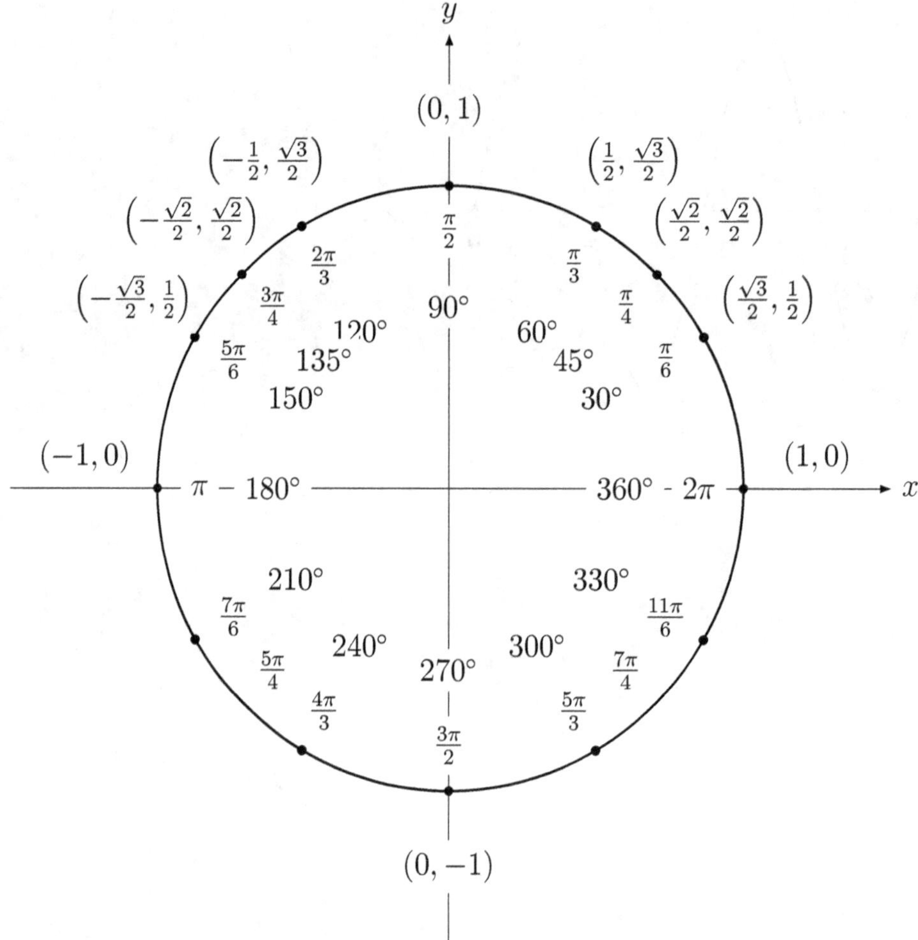

You may use the figure above as reference to help learn the relationship between angles in degrees and radians.

It shows the coordinates of important points on the unit circle, and the relative angles in radians and degrees. For the test, the main takeaway is that the conversion $\pi = 180°$. Maybe it helps to remember that π is the ration of the circumference to the diameter.

9.3　Exercises

Answers are in the back of the book, Appendix C.

1. What is the ratio of the area of the larger circle to the area of the smaller circle such that the radius of the larger circle is three times the radius of the smaller circle?

2. Given the three tangent circles of unit radius, find the area in the center of the triangle, excluding the circles.

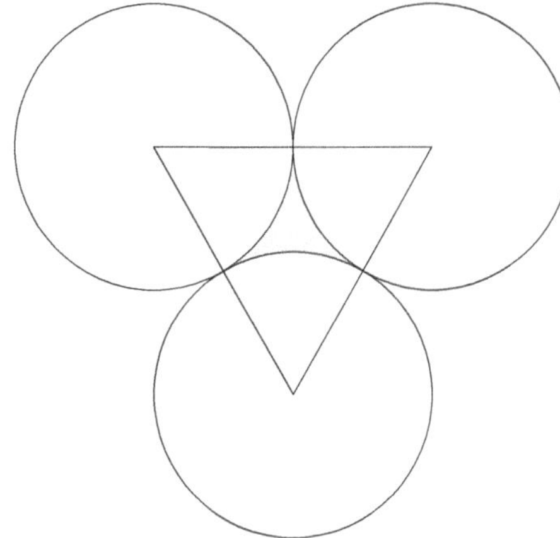

3. Find the area of the four curves sections in the lower triangle, i.e. the triangle minus the sections of the circles.

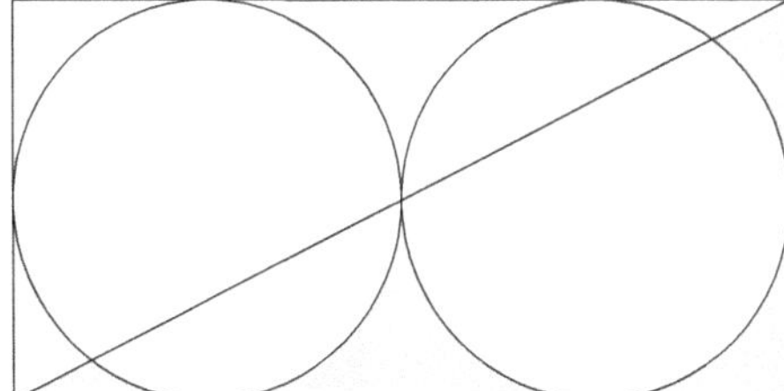

10

Complex Numbers

10.1 So-called Imaginary

Usually there are only a couple of problems dealing with complex numbers in the test.

Complex numbers are real numbers plus and imaginary numbers. Unfortunately the name "imaginary" is rather misleading since they are no more imaginary than $\sqrt{3}$ or -133. Even the great mathematician Gauss had suggested calling them "lateral" numbers, but that didn't catch on.

Just as negative numbers were introduced to solve problems such as $3 - 14$, for which there were no positive integer solutions, and rational numbers were introduced to solve problems such as $x^2 = 2$, so complex numbers were introduced to solve equations such as $x^2 + 16 = 0$.

If you have not been taught complex numbers yet you can simply follow the explanation that follows. The problems in the sat will usually be a simple operation: addition, subtraction, multiplication, or division.

95

We define i as $i^2 = -1$, in other words $\sqrt{-1} = i$. The complex number z is equal to $a + bi$, where a and b are real numbers and $i^2 = -1$.

On the test they will usually remind you that $i^2 = -1$

Addition and subtraction are relatively straightforward, we add the real parts and the imaginary parts.

$$(3 + 4i) + (2 + 5i) = (3 + 2) + (4 + 5)i = 5 + 9i$$

Multiplication is done using the distributive property. (Similar to multiplying two binomials.)

$$(a + bi) \times c + di = ac + adi + bci + bdi^2 = (ac - bd) + (ad + bc)i$$

Division takes a few more steps. First we will need to know the complex conjugate of a complex number is the same pair with the opposite operation for the imaginary part. $a + bi = a - bi$.

Then we use the fact that $\frac{a-bi}{a-bi} = 1$.

We can perform division using the complex conjugate of the denominator:

$$\frac{a + bi}{c + di} = \frac{a + bi}{c + di} \cdot \frac{c - di}{c - di} = \frac{(ac + bd) + (bc + ad)i}{c^2 + d^2} = \frac{ac + bd}{c^2 + d^2} + \frac{bc + ad}{c^2 + d^2}i$$

If you have been studied complex numbers already, you probably have been introduced to the complex plane. The Real numbers are along the x axis, similar to the the Cartesian plane or Coordinate system, and the Imaginary axis corresponds to the y axis in the Cartesian plane.

As you will discover later on in your studies you can actually think of the Imaginary axis as a third axis perpendicular to the xy plane. But that is sufficient for now.

10.2 Exercises

Answers are in the back of the book, Appendix C.

1. What is the following in the form $a + bi$?

$$\frac{15 + 5i}{3 - 2i}$$

2. What is $(3i)^4$?

11

Word Problems

11.1 Words into Equations

Often the difficulty in word problems, such as rate problems or percentages, is not simply the math, but how to interpret the problem. In other words you need to start from a given written paragraph and transform it into an algebraic equation, with no graphs, formulas, or diagrams to help translate the problem from English to Math.

Not all word problems in the test will ask for a numeric solution. Sometimes they will just ask which is the appropriate equation posed by the problem, or what a certain variable means in relation to the problem. So it is also a question of language.

First of all observe the words used in word problems.

If I have three pen and buy three more, how many do I have? The word "more" tells us we have a sum. Similarly with the word "and". *Mary has three books and five pencils, how many items does she have?* Other key words that indicate addition may be: *sum, total, increased by.* A subtraction may be called for when you come across the words: *difference, less, fewer,* or *decreased.*

Multiplication may be indicated by *times,* or *product.* Multiplication by 2 would be indicated by *twice* and division by *out of, per,* or *quotient.*

A comparison that indicates an equality may use such words as *as many as* or *as much as,* or the words *is, will be,* or *the same as.* Inequalities may use *more than* or *less than.*

Rates

Rate problems are some quantity defined by a ratio of two other quantities. For example, speed is a rate defined by the distance divided by the time required to cover that distance.

$$V = \frac{\text{distance}}{\text{time}}$$

The rate at which work is done is, similarly,

$$W = \frac{\text{output}}{\text{time}}$$

Of course, by manipulating the equations you can solve if the unknowns is one of the elements of the ratio.

Perhaps the best way to understand how to tackle word problems is by looking at some examples, and follow the solution step by step.

To solve a word problem one must understand what kind of solution is being sought: is it a total number, a fraction, a speed or how many objects will be produced in one day? And then what are the elements that can lead to the solution. Look at the units of measure: is it kilometers per hour, a ratio, or a percentage discount?

Let us look at a concrete example. Suppose that 250 wild geese were banded with identification tags. They were then released in the wild where they joined the flocks migrating north. At a lake in Canada biologists counted about 1500 geese of which 25 had the identifying bands. Estimate the total number of geese that migrated[1].

1. What is being asked? To find the total number T.

2. What other information do we have? 250 geese tagged, and 12:1500 is a ratio we know.

3. The geese at the lake are only part of the total.

4. The ratio of tagged to total must equal ratio of tagged at the lake to all the ones at the lake.

5. So 25:1500 = 250:T

6. $\frac{25}{1500} = \frac{250}{T} \rightarrow T = \frac{250 \cdot 1500}{25} = 15,000$

Let's try another example:

The price of a pair of shoes has increased by 15% because of inflation. The new price is 123 USD. What was the original price?

1. What is being asked? To find the original (lower) price P.

2. What other information do we have? P increased by 15% and the new price is 123 USD.

3. We know that $P \cdot 0.15$ is what was added to P to make 123.

4. That can be written as $P + 0.15 \cdot P = 123$

5. In other words $P(1 + 0.15) = P(1.15) = 123$

---

[1]There are a number of assumptions we do not specify such as all geese survive, the ratio at the lake is representative etc. In real life biology population estimates this is called Mark and Recapture, and all the assumptions must be stated.

6. $P = \frac{123}{1.15} = 106.96$

7. Quick check for plausibility, P is less than the new price so that looks good.

It takes three minutes to print a 10 page brochure on the new HX printer. Can you write that as an equation of brochures to minutes?

1. What is the problem asking? An equation that expresses a relation. So it is a rate.

2. How many minutes to brochures (m) to (b).

3. $\frac{3m}{10b} = \frac{xm}{1b} \rightarrow \frac{3}{10} = \frac{m}{b}$

4. $m = \frac{3}{10}b$ (Notice that this is a slope in the xy coordinate system)

12

Permutations, Combinations

12.1 Counting

A combination is a selection of objects in which order does not matter: a combination of ingredients in a salad, a group of people to form a committee.

If the order matters, it is a permutation.

Some examples are: the possible ways to arrange some books on a shelf, the possible way we could line up a group of children, if three people are competing, the possible order in which they could come.

So technically the three digit lock on briefcases and suitcases are not really combination locks because the order of the digits is important.

Of combinations and permutations there are different types.

1. Permutation without repetition, $n \cdot (n-1) \cdot (n-2) \ldots 2 \cdot 1 \to n!$

For example: In the painting *Last Supper* by Leonardo, all 13 figures are arranged in a row. Assuming he could have portrayed them in any order, how many permutations would there be?

By reasoning, we know that there are 13 possibilities for the first figure, once that one has been selected there are 12 possibilities for the second, that makes $13 \cdot 12$ possibilities for the two slots alone.

Continuing this way we see that there are $13 \cdot 12 \cdot 11 \cdot 10 \cdot 9 \cdot 8 \cdot 7 \cdot 6 \cdot 5 \cdot 4 \cdot 3 \cdot 2 \cdot 1$ possibilities (which is a lot).

Because such a multiplication is quite common we will use the special notation, $n!$ called n factorial.

The definition of factorial is

$$n! = n \cdot (n-1) \cdot (n-2) \cdot \ldots \cdot 1$$

Note that as n increases, the result grows quite rapidly.

n		
1	$1! =$	1
2	$2! =$	2
3	$3! =$	6
4	$4! =$	24
5	$5! =$	120
6	$6! =$	720
7	$7! =$	5040
8	$8! =$	40,320
9	$9! =$	362,880
10	$10! =$	3,628,800

So any problems were the order is important, and all arrangements of all the n elements are possible of the set, the total resulting permutations will be given by $n!$

2. Permutations with repetitions

Let us take, for example, the lock which we saw was incorrectly called a combination lock. This one is a permutation with repetition, because, for example, if we set 7 as the first number, we could still set a 7 in the second slot, and the third, so the three numbers could be set to "777".

We proceed the same way as before. Since there are 10 digits (assuming 0 is also possible), there are 10 possibilities for the first slot. There are 10 possibilities for the second, and 10 for the third. So in all there are 10^3 possible permutations

We then define the very simple formula for permutations with repetitions as:

$$n^r = n \cdot n \cdot n \ldots \ldots \cdot n$$

3. Permutation without repetition of a subset.

For example: In a swimming competition there are 10 contestants. first, second and third winners in a competition. They are all equally good, so we want to know how many possible line ups there could be.

Again, by reasoning we see that there are 10 possible winners for first place, which leaves 9 possible winners, once first place has been picked, leaving 8 for third position. So there are $10 \cdot 9 \cdot 8$ possible line ups.

But how do we represent this mathematically, that is, with a formula? What we do is "divide out" the tail end we don't need, $(10 - 3)! = 7!$

$$\frac{(10 \cdot 9 \cdot 8) \cdot (7 \cdot 6 \cdot 5 \cdot 4 \cdot 3 \cdot 2 \cdot 1)}{7 \cdot 6 \cdot 5 \cdot 4 \cdot 3 \cdot 2 \cdot 1}$$

which gives us the permutation of possible first, second, and third place winners. And with this reasoning we can write the general formula for k out of n elements as

$$\binom{n}{k} = \frac{n!}{(n - k)!}$$

Note: many write $_nC_k$ or ^nC_k instead of $\binom{n}{k}$.

4. Combinations without repetitions of k out n elements:

Now, imagine that Jane wants to choose three toys from the ten to give to her dog. The dog obviously does not care in what order she picks them.

If we proceed as with the permutation of k out n we would get a number that is too big, because it would count each order of the three chosen items as being different.

In a sense we have to "collapse" that number so that, for example $\{A, B, C\} = \{B, A, C\}\ \{C, B, A\}\ldots$ In other words the 6 possible permutations of the three object have to be counted as only 1. But the 6 is given by the possible permutations of 3 objects, $3!$.

So to obtain the general formula, we divide the permutations we got by the number of object k factorial, that gives us the combination of k objects out of n

$$^{n}P_{k} = \frac{n!}{k!(n-k)!}$$

Note: many write $_{n}P_{r}$ or $P(n, k)$ instead of $^{n}P_{r}$.

5. Combinations of a subset with repetition.

For example, if you get an ice-cream and pay for 3 scoops. You have 5 flavors to choose from, vanilla, lemon, strawberry chocolate and pistachio. And you are allowed to repeat a flavor with the other scoops.

This would be $n = 5$ flavors of ice cream, choose $k = 3$ scoops.

Therefore the possible choices could be (vanilla, vanilla, vanilla) or (vanilla, chocolate, pistachio) or (strawberry, lemon, strawberry) and so on.

Now, because strawberry, lemon, strawberry and strawberry, strawberry, lemon and lemon, strawberry, strawberry are all considered the same we see that it cannot be 5^{3}, but something smaller.

The formula is

$$\frac{(k + n - 1)!}{k!(n - 1)!}$$

lets tackle a smaller set first.

Let's try combination 2 out of 3. Let's use the three letters (A, B, C).

There are 3 possibilities for 2 being the same. There are 2 possible choices for each of the three letters, for example { A, B} or {A, C}.

A	A
B	B
C	C
A	C
B	A
B	C

For a total of 6 combinations.

Using the formula also gives us 6, as it should be.

$$\frac{(3 + 2 - 1)!}{2!(3 - 1)!} = \frac{4 \cdot 3 \cdot 2}{2 \cdot 2} = 6$$

Let's roll up our sleeves and try 3 out of 5. We'll use the letters (A, B, C, D, E) instead of ice cream flavors to make things easier.

There are 5 ways to have all 3 the same, obviously: (A, A, A), (B, B , B) etc.

For each of the 5 flavors there are 4 ways to have 2 be the same, for example, (A, B, B), A (A, C, C), (A, D, D), (A, E, E), and so on. That makes 20 combinations.

For all three flavors to be different we can use the formula for combinations without repetition which gives 10 combinations.

The total then seems to be $5 + 20 + 10 = 35$.

With the formula we have

$$\frac{(5 + 3 - 1)!}{3!(5 - 1)!} = \frac{7!}{3!4!} = \frac{7 \cdot 6 \cdot 5 \cdot 4 \cdot 3 \cdot 2}{3 \cdot 2 \cdot 4 \cdot 3 \cdot 2} = 35$$

12.2   Probability

Probability theory is a logical analysis of the likelihood of an event occurring. If the event cannot occur, in other words it is impossible, then it is indicated by zero. If an outcome is certain then it is indicated by one.

The probability of an event E can then be expressed as $0 \leq P(E) \leq 1$.

Between those two extreme values are the expected values.

With discrete events, that is separate possible events, for example heads or tails of a coin toss, or the numbers 1 through 6 of the roll of a die, or picking a specific card out of a deck of 52, the probability is calculated as the number of ways the outcome can occur divided by all possible outcomes.

For example, there are 4 kings in a deck of cards, so the probability of randomly selecting a king from a set of 52 cards would be 4/52 or 1/13.

The chances of rolling an odd number with the roll of a die is $3/6 = 1/2$. As a formula, then, probability is the ways an event can occur divided by the total number of outcomes, technically called the sample space.

$$P(E) = \frac{\text{ways it can happen}}{\text{sample space}}$$

The probability of tossing heads with a fair coin we can express as $P(H) = 1/2 = 0.5$. Note we can write probability as fraction or a decimal.

There are more possible number outcomes with more coins of course. With two coins you could have HH (two heads), (TT), two tails, (TH), or (HT), that is four total possible outcomes.

The probability of two or more independent events is the product of the probabilities. The probability of tossing one head and one tail with two coins would be $1/2 \cdot 1/2 = 1/4$

What is the probability of three heads and three tails with six coins?

All possible outcomes is $2^6 = 64$. That is the product of total outcomes for each coin.

There is only one way of getting all heads, six possible ways of getting one head. The same goes for no tails and one tail.

To calculate how many ways we could have two heads we can start assuming the first coin is heads, that means there are five ways one of the other coins can be heads.

Similarly, if the second coin is heads, one of the other four could be heads. And so the third coin leaves three possible heads, the fourth two, and only one ways for the last two to be heads. Symmetrically the same calculations for the tails.

Adding them up, for the heads we get $1 + 6 + +5 + 4 + 3 + 2 + 1 = 22$, and the same for tails for a total of 44 possible outcomes. But we wanted to calculate 3 heads and 3 tails. Since the probability of an event is between 0 and 1, then the probability of the complement of that event, that is, it not occurring is $1 - P(E)$.

So in this case the probability of not getting three heads and three tails is $44/64$, So the opposite, would be $64 - 44 = 20$ and the probability is then $22/64 = 5/16$

That is for independent events. The result of tossing one coin does not affect the outcome of the others.

The probability of independent events is then

$$P(A) \ \text{AND} \ P(B) = P(A) \cdot P(B)$$

Let us assume we are drawing marbles out of an urn. There are red, blue, and green marbles. If we extract as marble, note the color and put it back, it will have no influence on the next extraction (assuming we mix them again). But if we don't put the marble back, that will certainly change the probability of the next extraction. These then are dependent events.

Assume there are 2 marbles of each of the three colors. The probability of selecting a red marble on the second draw will be different if the first draw I picked a red marble or one of the other two colors. In the first case, in the second draw I will have one red marble out of 5 left. In the second case I will have 2 red out of the five in the urn.

The probability of two events that are mutually exclusive, that is one OR the other can occur is just the sum of the two events. For example the probability or rolling a five or a six with a die is $P(5 \ \text{OR} \ 6) = P(5) + P(6) = 1/6 + 1/6 = 1/3$.

The probability of events A AND B both occurring is expressed as the intersection of the two events, $PA \cap P(B)$, and if the two events are mutually exclusive then $P(A) \cap P(B) = 0$

Now if we want to calculate the probability of two dependent events we have to specify whether we want to know the probability of both (the first AND the second) or the probability of either event (one OR the other).

For example, the probability of drawing two kings from a deck of cards. The probability of a king in the first draw is $P(K_1) = 4/52 = 1/13$. Having drawn a king the probability of drawing a second king is now $P(K_2) = 3/51 = 1/17$. Then $P(K_1) \ \text{AND} \ P(K_2) = 1/13 \cdot 1/17 = 1/221$

Another way of expressing the probability of two dependent events both occurring is

$$P(A \text{ AND } B) = P(A) \cdot P(B|A)$$

The term $P(B|A)$ is read as the probability of event B given that event A occurred.

The probability of either of two events occurring is $P(A)$ OR $P(B)$, and it is the sum of the two events minus the intersection of the two events (they both occurring, because otherwise we count it twice, just like when working with sets).

$$P(A) \text{ OR } P(B) = P(A) + P(B) - P(A \cap B)$$

12.3   Exercises

Answers are in the back of the book, Appendix C.

1. Is creating an 8 letter password a combination or permutation?

2. Is choosing a Chairman and a Co-Chairman out of 15 people a combination or a permutation?

3. Out of five semifinalists, Alice, Bob, Carol, Danny and Elsa, three will be selected as finalists. If the selection is random, how many ways of choosing three people are there?

4. Out of four contestants three will be randomly select for first, second and third prize. How many ways of selecting the winners are there?

5. What is the possibility of getting two heads in a row out of three coin tosses given that two heads occur?

6. We need to form teams of 5 each side team in a class of 12 students. How many different teams can be formed?

7. How many 4 digit numbers can we make using the digits 3, 6, 7 and 8 without repetitions?

8. How many 3 digit numbers can we make using the digits 2, 3, 4, 5, and 6 without repetitions?

9. How many 6 letter words can we make using the letters in the word LIBERTY without repetitions?

10. In how many ways can you arrange 5 different books on a shelf?

11. In how many ways can you select a committee of 3 students out of 10 students?

12. How many triangles can you make using 6 non collinear points on a plane?

13. A committee including 3 boys and 4 girls is to be formed from a group of 10 boys and 12 girls. How many different committee can be formed from the group?

14. In a certain country, the car number plate is formed by 4 digits from the digits 1, 2, 3, 4, 5, 6, 7, 8 and 9 followed by 3 letters from the alphabet. How many number plates can be formed if neither the digits nor the letters are repeated?

15. There are 5 doors to a lecture room. In how many ways can a student enter the room through a door and leave the room by a different door?

16. In how many ways can 3 students be selected from a group of 12 students to represent a school in the inter school essay competition?

17. (Difficult) How many squares are there in a chess board?

13

Statistics

13.1 Average, Median, Mode

The average, or the mean, is found by adding the values of the elements, and dividing the total by the number of elements. So if the grades of seven students are: {5, 6, 7, 8, 9, 9, 10 }, then the average is $54/7 = 7.7$).

The median is {5,6,7,8,9,9,10} the value in the middle when the values are placed in order. In this case it's the fourth element, 8.

If there are an even number of elements in the list, for example, in our list, if there had been another element,such as 12, giving us {5,6,7,8,9,9,10, 12} then we find the average of the two numbers in the middle, that is, added and divided by two, in this case giving 8.5.

The weighed average means that the differing averages are summed proportional to the number of items in each of the averages divided by the total.

In the unlikely event you get a weighed average, for example, a measure of the length of the same species of small toads in two different lakes. One lake, 50 toads were measured for an average of 51 mm, and in the second lake 25 toads were measured with an average of 45 mm. The average of the toads for the two lakes then is

$$\frac{25}{75} \cdot 45 + \frac{50}{75} \cdot 51 = \frac{1}{3} \cdot 45 + \frac{2}{3} \cdot 51 = 15 + 34 = 49$$

The formula in general is

$$\frac{n_1}{T} \cdot a_1 + \frac{n_2}{T} \cdot a_2$$

where n_1 and n_2 are the number of items respectively in set 1 and set 2, the total is $T = n_1 + n_2$, and a_1 and a_2 are the averages, respectively. for the two sets.

This can be extended for any number of sets

$$\frac{n_1}{T} \cdot a_1 + \frac{n_2}{T} \cdot a_2 + \frac{n_3}{T} \cdot a_3 + \frac{n_4}{T} \cdot a_4 \ldots$$

The mode is the most frequent value, here 9. If the list had, for example, 5,6,7,7,9,9,10 two numbers with the same highest frequency they are both modes, such as here 7 and 9.

Standard deviation

The standard deviation gives an idea how "spread out" the values are. That is, it tells us whether the values tend to bunch up very close to the mean or they are further apart.

A normal curve is the graph that shows this "spread". The normal curve means that about 34% of the values are within one standard deviation above the mean, and another 34% are within one standard deviation below the mean. So 68 % of the values are underneath the central part of the "bell curve".

A remaining 14 % are within 1 and 2 standard deviations above, and similarly, 14 % between 1 and 2 standard deviations below the mean. In the tail are the remaining 2 % greater than 2 standards deviations, and 2 % less than 2 standard deviations from the mean.

You can usually do with remembering the percentages as 34-14-2. These are approximate values. If more precise values are required they will supply them in the test.

The usual notation for standard deviation is sigma, σ and the subscript tells whether it's 1 or two standard deviations from the mean, σ_1 or σ_2, and the sign, positive or negative whether greater or less, respectively, than the mean.

In the graph below, the percentages above the horizontal bars gives the total for the combined first and second standard deviations around the mean.

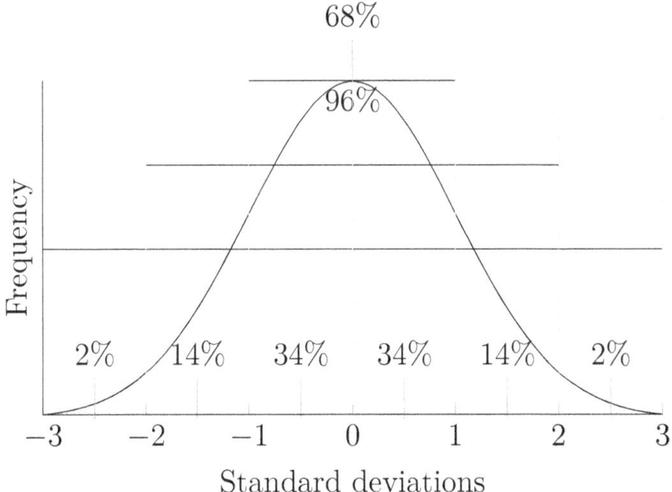

So in the graph above, σ_1 will have the lowest deviation from the mean, and σ_3, the highest.

You may also note that at each standard deviation is an inflection point of the curve.

Properties

- A normal curve is symmetrical about the mean μ.

- The mean divides the curve into two equal halves.

- The total area under the curve equals 1.

- Inflection points at $\mu+\sigma$ and $\mu-\sigma$. In other words, one standard deviation above and one below the mean, respectively.

- Only two parameters are necessary, μ and σ since the curve is completely determined by them: the mean μ and the standard deviation σ or variance σ^2.

Let's compare two normalized curves.

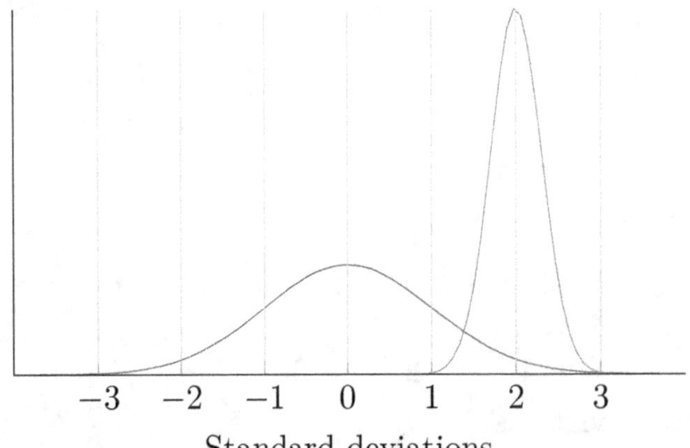

Notice that the flatter curve is also is wider than the tall bell curve. That means its standard deviation is greater. We may also notice that the mean or average for the tall one is greater than that of the flatter curve. Its standard deviations are not shown, but both its σ_1 and σ_2 would be less than one standard deviation of the flatter curve.

13.2   Box and Whiskers

The Box and Whisker plot is a graphical representation of the raw data showing five significant values: the Minimum is the lowest data point and the Maximum, the greatest data value, the Median of the set of data, the First quartile (Q1) or lower quartile is calculated as if it were the median of the lower half of the set of data, that is it is between the minimum and the Median of the data set the Third quartile (Q3) or upper quartile, is the median of the upper half of the data set, and, lastly, the Interquartile Range (IQR): is the distance between the upper and lower quartile.

In statistics, the percentile indicates the value below which a given percentage of observations in a group of observations falls. For example, the 60th percentile is the value (or score) below which are 60% of the data and 40% will be above the 60th percentile.

The percentile rank of a score is the percentage of values that are equal to or lower than it. For example, a test score that is greater than 75% of the scores of people taking the test is said to be at the 75th percentile, or a percentile rank of 75.

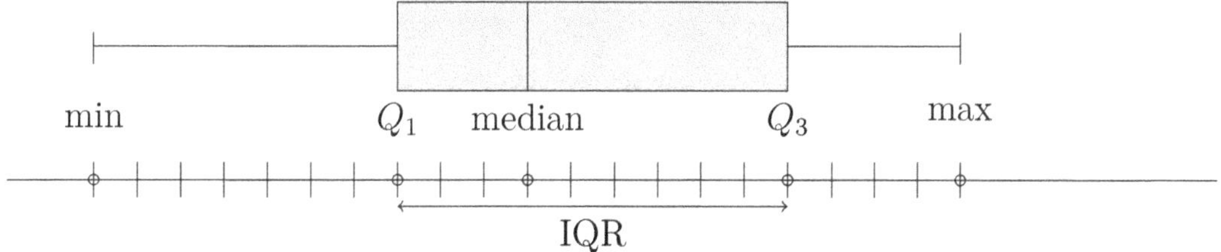

Extra material 1: standard deviation

These two small subsections go into the two concepts of standard deviation and correlation a little bit deeper. They are extra material.

Here is a bell curve showing more precise percentages, and showing that above 3 standard deviations there is a tail of 0.13%, as well as below, which we did not take into account in our approximations.

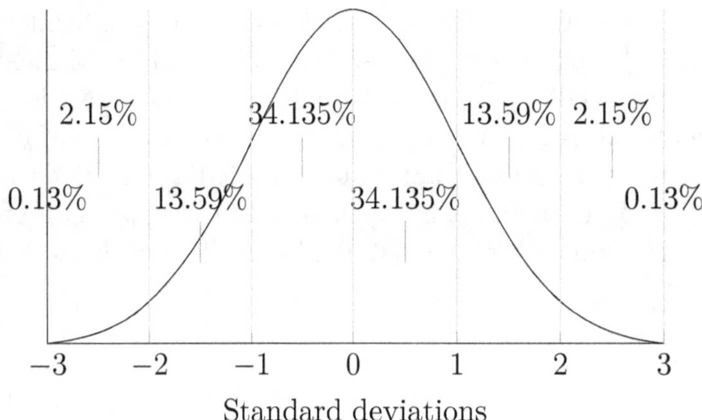

If you want to know how the standard deviation is calculated , where $\bar{x}$ is the average, and x is the variable.

$$S_x = \sqrt{\frac{\sum(x - \bar{x})^2}{n}}$$

These are included for completeness, you are not expected to know them.

For those who want to really delve a little deeper.

By way of example, we can standardize the normal curve by making the mean equal 0 and the standard deviation equal 1 using the following transformation:

$$Z = \frac{X - \mu}{\sigma}$$

For example the normal curve we saw above where $\mu = 2$ and $\sigma = 1/3$ can be normalized as follows:

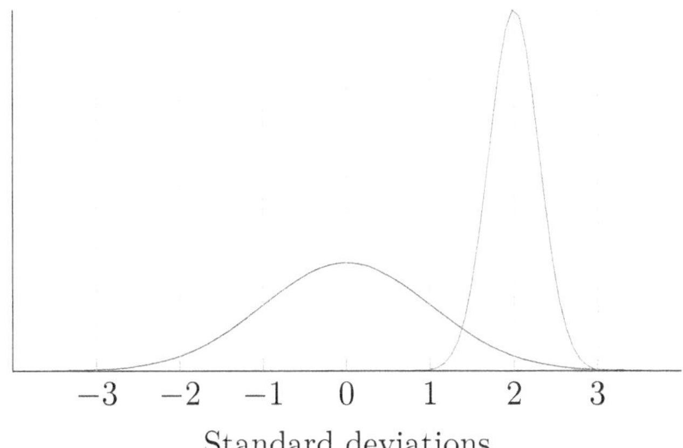

Standard deviations

Though the two graphs have different μ and σ, they have the same area.

The values between x_1 and x_2 will the same area as the area in the normalized curve between z_1 and z_2, so we have the equivalent probabilities:

$$P(x_1 < X < x_2) = P(z_1 < Z < z_2)$$

So in our example, $\mu = 2$ and $\sigma = 1/3$ the area between $1/2$ standard deviation to the right and 2 standard deviations to the right will be between

$$\mu + \left(\frac{1}{2} \cdot \sigma\right) = 2 + \frac{1}{2 \cdot 3} = \frac{13}{6} \text{ and } \mu + 2 \cdot \sigma = \frac{8}{3}$$

This is equal to the area under the normal curve between

$$z_1 = 0.5 \text{ and } z_2 = 2$$

Extra material: Proportional Breakdown

If a continuous variable X is normally distributed with mean μ and standard deviation σ, we write

$$X \sim N(\mu, \sigma^2)$$

Extra material: Correlation

A visual means of assessing correlation is the scatter plot with the independent variable along the x-axis and the dependent variable along the vertical y-axis.

If the dependent variable increases as the independent variable increases then we can say there is a positive correlation. If the dependent variable decreases as the independent variable increases then there is a negative correlation.

We can calculate a coefficient (called Pearson Correlation Coefficient) r that measures the strength of the correlation. The formula for r is S_{xy}, the co-variance of x and y, divided by the product of the standard deviation of x and the standard deviation of y, $S_x S_y$.

$$r = \frac{S_{xy}}{S_x S_y}$$

13.3   Exercises

Answers are in the back of the book, Appendix C.

1. The average (arithmetic mean) of x and y is 30. If $z = 10$, what is the average of x, y, and z?

2. If the average (arithmetic mean) of 5 consecutive integers is 11, what is the sum of the least and greatest of the 5 integers?

3. If w is the average of the numbers a, b, c and d, then what is the average of m(a + k), m(b + k), m(c + k) and m(d + k)?

4. The standard deviation on a test was 12 points, and the mean was 70. If the scores fell along a normal distribution and student X scored 95 points, then student X scored higher than approximately what percent of students?

5. In a survey to determine how many bottles of drink X are consumed by 100,000 people a year. From the data the mean 29 is approximately normally distributed with a deviation of 4 per person. How many of the surveyed people drank more than 25 in one year?

A

Appendix

A.1   Further Reading

Most of these books are aimed at non mathematicians. I hope that reading one or more may make you show a more interesting side of math and maybe greater inspiration.

They may not only give insight into what mathematics is all about, but provide greater appreciation of mathematics far beyond what you may have learned in school.

Richard P. Feynman , Ralph Leighton *Surely You're Joking, Mr. Feynman!: Adventures of a Curious Character*

W. W. Norton & Company, Reissue edition, 2018. There is no mathematics in this delightful biography of one of the greatest physicists, and brilliant mathematician, of the last century. Funny and inspiring. His biography was made into the 1996 film "Infinity".

Margot Lee Shetterly *Hidden Figures: The American Dream and the Untold*

Story of the Black Women Mathematicians Who Helped Win the Space Race, William Morrow Paperbacks; Media Tie In edition, 2016.

Another book without math, but about mathematicians. If you haven't seen the 2016 movie of the same name, check it out. This is the story of women mathematicians who had to fight great odds and hostility, whose contribution has only recently been acknowledged.

R. Kanigal *The Man Who Knew Infinity*, Abacus, 1992.

The third biographical book on the list. The very moving and interesting story of one of the most mysterious mathematicians of the modern age. An unschooled Indian clerk whose genius was recognized by England's greatest mathematicians. This also was made into a movie in 2015, with the same title

Richard Courant and Herbert Robbins *What is Mathematics: An Elementary Approach to Ideas and Methods* Oxford Univ. Press (1941) 1978.

This is one of the books that made me decide to take up mathematics. It is a very different, yet rigorous approach from school textbooks. The authors try to provide an understanding mathematics by starting from what numbers are on through calculus. It goes far deeper into the meaning of many of the things you may have learned in school.

Philip J. Davis and Reuben Hersh *The Mathematical Experience* Penguin Books 1986. This is also for non mathematicians. This takes a look at the modern practice of mathematics, with insights into what mathematicians do and who they are.

William Dunham *Journey Through Genius: The Great Theorems of Mathematics* Wiley Science ed. 1990. This is enjoyable reading historical look at some of greatest mathematical theorems for the layman.

William Dunham *The Mathematical Universe: An Alphabetical Journey Through the Great Proofs, Problems, and Personalities* Wiley Science ed. 1994

Different mathematical topics presented for the non-mathematicians, informal yet solid foundations, with interesting biographies of mathematicians.

Eli Maor *To Infinity and Beyond: A Cultural History of the Infinite* Princeton Univ. Press 1991.

Explores many of the lesser known ideas about infinity. It has many illustrations and discusses the impact of the concept of infinity on art, culture, and on the other sciences.

Ed. by J. H. Ewing *Numbers* Springer-Verlag (GTM) 1990.

This one is more technical than the others, it contains different accounts of what numbers are, from ancient Egypt to today. If you want to delve deeper into the realm of numbers you will find this challenge rewarding.

E. T. Bell *Men of Mathematics* Touchstone, Reissue edition, 1986. This is a very enjoyable overview of the work of major mathematicians, though the biographies are considered outdated and inaccurate in some ways.

B.1   A Bit About Number Systems

This section is for those who wish to get a peek at a slightly more formal approach to numbers and their properties. It is adapted from an article I had published online years ago.

I will give a brief explanation of the following concepts:

1. $\mathbb{N}$, the set of natural numbers, is a monoid

2. $\mathbb{Z}$, the set of integers, is an integral domain

3. $\mathbb{Q}$, the set of rationals, is a field

4. in the field $\mathbb{R}$, the set of reals, the order is complete

5. the field $\mathbb{C}$, the set of complex numbers, is algebraically complete

If you were asked by a young child to give them arithmetic problems, so they could show off their newly learned skills in addition and subtraction, perhaps

127

after a few problems such as: $2 + 3$, $9 - 5$, $10 + 2$ and $6 - 4$, you could try them something a little more difficult: $4 - 7$ only to be told " That's not allowed."

What you may not realize is that you and the child did not just have different objects in mind, counting or natural numbers, $\mathbb{N}$, (that do not contain negative numbers) and integers, $\mathbb{Z}$, (that do). You are actually thinking of entirely different algebraic systems. In other words a set of objects (they could be natural numbers, integers or reals) and a set of *operations*, or rules regarding how the numbers can be combined.

We will take an informal tour of some algebraic systems, but before we define some of the terms, let us build a structure which will have some necessary properties for examples and counterexamples that will help us clarify some of the definitions.

We know that any number that is divided by six will either leave a remainder, or will be divided exactly (which is after all the remainder 0). Let us write any number by the remainder n it leaves after division by six, denoting it as $[n]$. This means that, 7, 55 and 1 will all be written $[1]$, which we call the class to which they all belong: i.e. 7 in $[1]$, 55 in $[1]$, or, a bit more technically, they are all equivalent to 1 modulo 6, written formally as 1 mod 6. The complete set of this class will contain six elements, that is called partitioning numbers into equivalent classes because it separates (or partitions) all of our numbers into these classes, and any one number in a class is equivalent to any other in the same class.

One interesting thing we can do with these classes is to try to add or to multiply them. What can $[1] + [3]$ mean? We can, rather naively try out what they mean in "normal" arithmetic: $[1] + [3] = [1 + 3] = [4]$. So far so good, let us try a second example 25 in $[1]$ and 45 in $[3]$, their sum is 70 which certainly belongs to $[4]$. Here we see what we meant above by equivalence, 25 is equivalent to 1 as far as this addition is concerned. Of course this is just one example, but fortunately it can be proven that the sum of two classes is always the class of the sums.

Now this is the kind of thing we all do when we add hours for example, 7 (o' clock) plus 6 hours is 1 (o' clock), and all we are really doing is adding hours (modulo 12).

The neat part comes with multiplication, as we will see later on. But for now just remember, it can be proven that something like $[4] \times [5] = [2]$ will work: the product of two classes is the class of the product.

Now for some of the necessary terminology: Monoids and Groups

To define a group let us take a set of objects and a rule (called a binary operation) which allows us to combine any two elements of this set. Addition is an example from math, or ANDing in some computer languages.

The set must be closed under the operation. That means that when two elements are combined the result must also be in the set. For example the set containing even numbers will always give us an even number when two elements are added together. But if we restrict ourselves to odd numbers, their sum is not an odd number and so we know right off the bat that the set of odd numbers and addition cannot constitute a group. Some books will consider closure in the definition of binary operation, and others add it as one of the requirements for a group along with the ones that follow below.

The set and the operation is called a group if the binary operation satisfies the following criteria:

- the operation is associative, which means it doesn't matter how you group the elements you are operating on, for example in our set of remainders: $[1] + ([3] + [4]) = ([1] + [3]) + [4]$

- there is an identity element, meaning: one of the elements combined with the others in the set doesn't change them in the least. For example the zero in addition, or the one in multiplication.

- every element has an inverse with respect to that operation. If you combine an element and its inverse you get the identity (of that operation) back.

(Be careful with this last one, -3 is the inverse of 3 in addition, since they give us 0 when added, but $1/3$ is the inverse of 3 with respect to multiplication, since $3 \times 1/3 = 1$, the identity under multiplication.)

So we can see that the set of natural numbers $\mathbb{N}$ (with the operation of addition) is not even a group, since there is no inverse for 5, for example. (In other words there is no natural number which added to 5 will give us zero.) And so the third rule for our operation is violated. But it still has some structure, even if it is not as rich as the ones we'll see later on.

Sets with an associative operation (the first condition above) are called semigroups, and if they also have an identity element (the second condition) then they are called monoids.

Our set of natural numbers under addition is then an example of a monoid, a structure that is not quite a group because it is missing the requirement that every element have an inverse under the operation (Which is why in first grade $4 - 7$ is not allowed.)

What about the set of integers, is it a group?

By itself this question is nonsensical. Why? Well, we have not mentioned under what operation. OK, let us say: the set of integers with addition.

Now, addition is associative, the zero does not change any number when added to it, and for every number n we can add $-n$ and get zero. So it's a group all right.

In fact it is a special kind of group. When we can perform the operation on the two elements in any order (e.g $a + b = b + a$) then the group is called *commutative*, or Abelian in honor of Abel. Not every operation is commutative, for example three minus two is certainly not the same as two minus three. Our set of integers under addition is then an Abelian group.

On to Rings. If we take an Abelian group (remember: a set with a binary operation) and we define a second operation on it we get a bit more of a structure than we had with just a group.

If the second operation is associative, and it is distributive over the first then we have a ring. Note that the second operation may not have an identity element, nor do we need to find an inverse for every element with respect to this second operation. As for what distributive means, intuitively it is what we do in math when perform the following change: $a \times (b + c) = (a \times b) + (a \times c)$.

If the second operation is also commutative then we have what is called a commutative ring. The set of integers (with addition and multiplication) is a commutative ring (with even an identity - called unit element - for multiplication).

Now let us go back to our set of remainders. What happens if we multiply $[5] \times [1]$? We see that we get $[5]$, in fact we can see a number of things according to our definitions above, $[5]$ is its own inverse, and $[1]$ is the multiplicative element. We can also show easily enough (by creating a complete multiplication table) that it is commutative. But notice that if we take $[3]$ and $[2]$, neither of which are equal to the class that the zero belongs to $[0]$, and we multiply them, we get $[3] \times [2] = [0]$. This bring us to the next definition. In a commutative ring, let us take an element which is not equal to zero and call it a. If we can find a non-zero element, say b that combined with a equals zero ($a \times b = 0$) then a is called a zero divisor.

A commutative ring is called an integral domain if it has no zero divisors. Well the set of integers $\mathbb{Z}$ with addition and multiplication fulfills all the necessary requirements, and so it is an integral domain. Notice that our set of remainders is not an integral domain, but we can build a similar set with remainders of division by five, for example, and voilÃă, we have an integral domain.

Let us take, for example, the set $\mathbb{Q}$ of rational numbers with addition and multiplication - I'll leave out the proof that it is a ring, but I think you should be able to verify it easily enough with the above definitions. But to give you a head start, notice the addition of rationals follow all the requirements for an Abelian group. If we remove the zero we will have another Abelian group, and that implies that we have something more than a ring, in fact, as we will see in the next section.

Fields

Now we can make one step further. If the elements of a ring, excluding the zero, form an Abelian group (with the second operation) then it is a field. For example, write the multiplication table of the remainders of division by 5, and you will see that it satisfies all the requirements for a group: (You will probably have noticed that the group does not contain the number five itself since $[5] = [0]$.)

	1	2	3	4
1	1	2	3	4
2	2	4	1	3
3	3	1	4	2
4	4	3	2	1

(Why isn't the set of divisors of six - excluding the zero and under multiplication - a group? That's easy enough, since we have excluded the zero we do not have the result of $[2] \times [3] = [0]$, which is not in our set, so it isn't closed.)

Ordering. Given a ring, we can say that it is ordered when you have a special subset of that ring behaves in a very special way. If any two elements of that special subset are added or multiplied their sum and their product are again in the special subset. Take the negative numbers in R , can they be that special subset? Well the sum seems to be alright, it is also a negative number. But things don't work with the product: it is positive. What about the positive numbers? Yep, and in fact we call that special subset, the set of positive elements. Now, we gave the definition for an ordered ring, we can also define an ordered field the same way.

But what does a complete ordered field mean? Well the definition looks rather nasty: it is complete if every non-empty subset which possesses an upper bound has a least upper bound.

Let's translate some of that, trying to lose as little information on the way.

A bound is something that guarantees that all of the elements of your set are on one side of it (reasonably enough). For example, certainly all negative reals are less than 100, so 100 is a bound (it is in fact an upper bound 'cause all negatives are "below" it). But there are lots of other bounds, 1, 5, 26 will all do nicely. The question now is, of all of these (upper bounds) which is the smallest, that is which one is "the border" so to speak? Does it always exists?

Look at the following numbers:

$$1.4 = 7/5, \ 1.41 = 141/100, \ 1.414 = 707/500, \ 1.4142 = 7071/5000, \ 1.41421 = 141421/100000\ldots$$

Now each of these is a rational number (it can be written as a fraction), and they are getting closer and closer to a number we've probably seen before (just take out your calculator and find the square root of two). So we can write the shorthand for this series as $\sqrt{(2)}$. Certainly we can find an upper bound for this series, 3 will do nicely, but so can 1.5, or 1.42. But what is the smallest. Well there isn't any. Not among the rational at least, because no matter what fraction you give I can give you one closer to the square root of two. What about the square root of two itself? Well it's not a rational number (I'll skip the proof, but it is really rather easy) so you can't use it. If you want another series which is really neat look up "Euler's formula."

And that is where the reals come in. Any set or reals that is bounded you can certainly find the smallest of these bounds. (By the way this "least upper bound" is abbreviated "l.u.b.", or "sup" for supremum.) We can also turn things around and talk of lower bounds, and of the largest of these etc. but most of that will be just a mirror image of what we have dealt with so far.

So that should be it. And for years that did seem to be it, we seemed to have all the numbers we'd ever care to have.

There was just one small stick in the works, but most people just sort of pretended not to notice, and that was that not all polynomials had solutions. One simple polynomial of this kind is $x^2 + 1 = 0$. It's so simple, yet there's no self respecting "real" number that would solve this polynomial. There were these funny answers which seemed like they should be solutions but no one could make any sense out of them, so they were considered imaginary solutions. Which was really too bad because they were given the name of imaginary numbers and now that the name stuck we realize that they are numbers just as good as any of the ones we have been using for centuries. And in fact that takes us to the last great pinnacle in this short excursion. The field of complex numbers.

We can define an algebraically closed field as a field where every non-constant polynomial (i.e. one with an x in it from high school days) has a zero in the field. Whew! This in short means that as long as the polynomial is not a constant number (which is no fun anyways) but something which looks like it wants a solution, like $5x^3 - 2x^2 + 6 = 0$ it will always have one, if you are working with complex numbers and not just reals.

There is another definition which is probably just as good, but may or may not be easier: A field is algebraically closed if every polynomial splits into linear factors. Linear factors are briefly factors not containing x to any power of two or higher, in other words in the form: $ax + b$. For example $x^2 + x - 6$ can be factored as $(x + 3)(x - 2)$, but if we are in the field of reals we cannot factor $x^2 + 1$, but we can in the field of complex numbers: $x^2 + 1 = (x - i)(x + i)$, where, you may recall, $i^2 = -1$.

Appendix C: Solutions

Chapter 2

1. **Solution:**

$$\sqrt{27} > \sqrt{25} \rightarrow \text{integer part} 5,\ 27 - 25 = 2 \text{numerator},\ 5 \cdot 2 = 10 \rightarrow 5.2$$

2. **Solution:**

$$\sqrt{138} > \sqrt{121} \rightarrow \text{integer part} 11,\ 138 - 121 = 17 \text{numerator},\ 11 \cdot 2 = 22 \rightarrow 11\frac{17}{22} \approx 11.77$$

3. **Solution:**

$$\{117, 144, 162, 198, 1017\} \text{ are divisible by } 9$$
$$\{144, 904, 1024\} \text{ are divisible by } 8$$

4. **Solution:**

$$771$$

5. **Solution:** The prime factors are $\{2, 5, 7, 13\}$ and the divisors are

$$\{2, 8, 10, 14, 20, 26, 28, 35, 50, 52, 65, 70, 91, 130, 140, 260\}$$

Chapter 3

1. **Solution:** The sequence will be the set $\{9, 12, 15, \ldots 72, 75\}$. Probably the fastest approach is to add 3 and 6 to the sequence. Then find the sum $\{3 + 6 + 9 + 12 \ldots + 75\}$. Factor out the 3 to get $3 \cdot \{1 + 2 + 3 + \ldots | 25\}$. Now calculate the sum, of the sequence, which is 325, multiply by 3. Don't forget to subtract the $3 + 6$ you added in the beginning.

$$966$$

2. **Solution:**
$$98^{\frac{1}{2}} = \sqrt{98} = \sqrt{49 \cdot 2} = \sqrt{49} \cdot \sqrt{2} = 7 \cdot \sqrt{2}$$

3. **Solution:** You could either calculate 130×0.15, or doing it piecewise. 15% is 10% + 5%, and 10% of 130 is 13, bur also. 5% is half of that, 6.5. So 10% + 5% is simply 13+6.5

$$19.5$$

Chapter 4

1. **Solution:** First solve for $g(5)$,

$$g(5) = 2 \cdot 5^2 - 5 = 2 \cdot 25 - 5 = 50 - 5 = 45$$

The solve for $f(45)$

$$f(45) = 45^3 + 3 \cdot 45 = 91,125 + 135 = 91,260$$

2. **Solution:** The first thing to note is that the numerator is not a perfect square, but it would be if we factor out the 2. Not also that we can also factor out the 2 in the denominator.

$$\frac{8x^2 - 50}{4x - 10} = \frac{2(4x^2 - 25)}{2(2x - 5)} = \frac{(2x + 5)(2x - 5)}{2x - 5} = 2x + 5$$

Chapter 5

1. **Solution:** The first thing is to write both equations in the form $m \cdot x + n \cdot y = p$, so we have $3x + 5y = 21$ and $2x - 10y = 14$, then, since we are asked to find x we can eliminate the y by multiplying the first equation by 2, and adding

$$6x + 10y = 42$$

and

$$2x - 10y = 14$$

we get $8x = 56$, which, by dividing by 8, we get $x = 7$

2. **Solution:** First rewrite the equation in the slope-intercept form (add $3x$ to both sides, and then divide by 5)

$$y = \frac{3}{5}x + 15$$

so the slope of a line perpendicular to the given line must have the negative inverse of the coefficient, therefore its slope must be

$$-\frac{5}{3}$$

Chapter 6

1. **Solution:** $x = -\frac{80}{d}$

2. **Solution:** $(2x + 3)(x + 2) \rightarrow x = -\frac{3}{2}, x = -2$

3. **Solution:** $\frac{3(x+1)}{\sqrt{2x+3}}$

4. **Solution:** $x \in (-1, 4)$

5. **Solution:** $[4, 7]$

6. **Solution:** $(5x + 3)(2x - 1)$

7. **Solution:** $\left(2 \cdot \sqrt{x} + 3 \cdot \sqrt{x}\right)^2 = (5\sqrt{x})^2 = 25x$ alternatively $(\sqrt{4x}+\sqrt{9x})^2 = 4x + 2(2\sqrt{x}3\sqrt{x}) + 9x = 25x$

Chapter 7

1. **Solution:** The percent can be written in decimal form as a 0.023, so the increase the first year is $1 + 0.023 = 1.023$. After m years the amount will be

$$2,200(1.023)^m$$

2. **Solution:** Solve by substituting the values in the equation

$$C(1 + p/100)^t$$

where C is initial amount, p is the percent interest. and t is the time elapsed (in years).

$$3,000(1.05)^3 = 3,472.88$$

Chapter 8

1. **Solution:** By extending the line to the other angle, we get the diagonal of the square, twice the given line segment. That is, $6\sqrt{2}$.

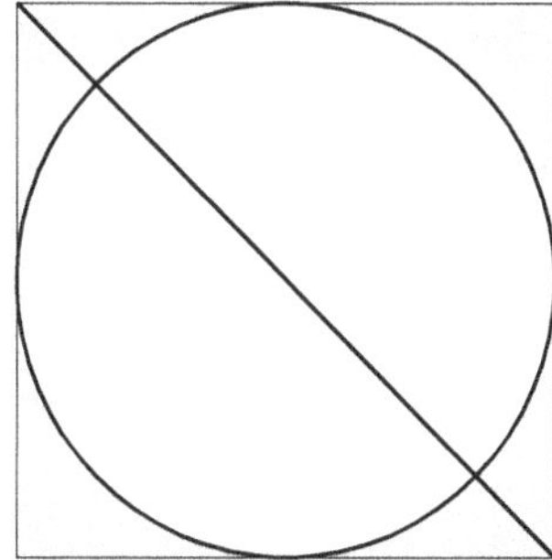

Since the ratio of the side of a square to its diagonal is $s\sqrt{2} : s$, the side of the square, and hence the diameter of the circle, must be 6. Therefore, by subtracting the area of the circle, A_c, from the area of the square, A_s, we get the desired remaining area.

$$A_s - A_c = 6^2 - \pi 3^2 = 36 - 9 \cdot \pi \approx 7.726$$

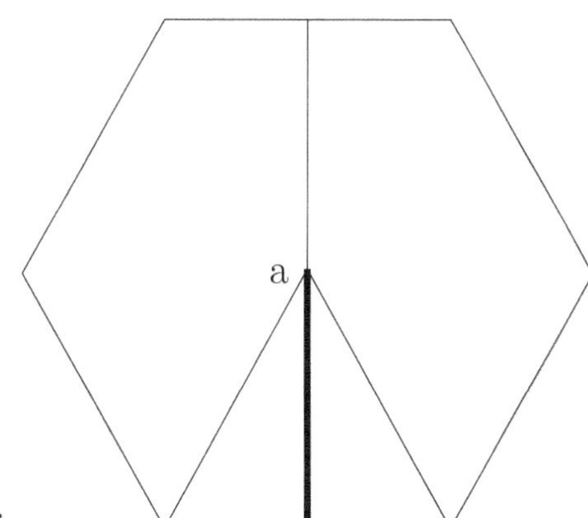

2. **Solution:**

   You can consider a hexagon as being composed of six equilateral triangles. The lower one being half of line segment a, so the height of each triangle is $3\sqrt{3}$ The base then is 6, and the area of each triangle is $\frac{3 \cdot 6\sqrt{3}}{2} = 9\sqrt{3}$, and there are 6 triangles in the hexagon:

$$6A = 6 \cdot \left(\frac{3 \cdot 6\sqrt{3}}{2} = 9\sqrt{3} \right) = 27\sqrt{3}$$

Chapter 9

1. **Solution:** Ratio of the areas is 9:1

2. **Solution:** The circles are tangent so the radii to the tangent points form an equilateral triangle of side 2. The area of the triangle can be found with the formula $\frac{\sqrt{3}}{4} \cdot s^2$ for an equilateral triangle, or, since half an equilateral triangle is s 30-60-90 triangle, the sides are $x, x\sqrt{3}, 2x$, in our case $x = \frac{1}{2}s1$. Both give us the area of the triangle $A_t = \sqrt{3}$ The sections of each circle is 60°, so each is a sixth and the three equal half a circle in area. The area of a unit circle is π, so the intersections add up to $A_c = \frac{\pi}{2}$ and so $A_c - A_t = \frac{\pi}{2} - \sqrt{3}$

3. **Solution:** The solution goes here

Chapter 10

1. **Solution:** Solve by using the inverse complex number $3 + 2i$ of the denominator.
$$\frac{15 + 5i}{3 - 2i} \cdot \frac{3 + 2i}{3 + 2i} = \frac{35 + 45i}{13} = \frac{35}{14} + \frac{45}{13}i$$

2. **Solution:** note the following pattern
$$i^1 = i, i^2 = -1, i^3 = -i, i^4 = 1, \text{and then repeats}$$
(Verify the pattern for yourself up to i^8). The solution then is
$$(3i)i^4 = 81 \cdot i^4 = 81$$

Chapter 12

1. **Solution:** Permutation

2. **Solution:** Combination

3. **Solution:** The possible combinations are ABC, ABD, ABE, ACD, ACE, ADE, BCD, BCE, BDE, CDE. Using the rule $_nC_r$ for $n = 5$ and $r = 3$ we get $\frac{5!}{(5-3)!3!} = 10$

4. **Solution:** This is a permutation, since the order is important. So we use the rule $_nP_r$ for $n = 4$ and $r = 3$ we get $\frac{4!}{(4-3)!} = 24$

5. **Solution:** Possibilities of two heads are HHT, HTH, THH, and two of them are heads in a row, so the probability is $2/3$

6. **Solution:** The order is not important so it is a combination problem. Hence the number of teams is given by $^{12}C_5$

7. **Solution:** $4!$

8. **Solution:** $^5P_3$

9. **Solution:** $^7P_6$

10. **Solution:** $5!$

11. **Solution:** $^{10}C_3$

12. **Solution:** $^6C_3$

13. **Solution:** $^{10}C_3 \cdot ^{12}C_4$

14. **Solution:** $^9P_4 \cdot ^{26}P_3$

15. **Solution:** 20

16. **Solution:** 220

17. **Solution:** 204

Chapter 13

1. **Solution:** If the mean is 30, then $x + y = 60$, so the mean with z added is $23\frac{1}{3}$

2. **Solution:** $11 = \frac{5n+10}{5}$ so $n = 9$ and the highest is 13, thus the sum is 22

3. **Solution:** $m(w + k)$

4. **Solution:** 95 is just two standard deviations above the mean, so the percentage is the sum of the percentiles below that: approx 97%

5. **Solution:** The mean is 29, with standard deviation of 4, so 68,000 people (68% of 100,000) drank between 25 and 33 bottles. And 16% are more than 1 standard deviation, so 16,000 drank more than 33. $16,000 + 68,000 = 84,000$ people drank more than 25 bottles.

Index

π, 83

absolute value, 24
algebraic expression, 35
angle, 43
arc, 84
arc of a circle, 83
associative property, 23
average, 113

binomials, 36

chess, 65
circle, center, 60
combination, 103
commutative property, 23
complex conjugate, 96
compound interest, 66
continued ratio, 32
correlation, 120
cube, 85

difference of two squares, 36
digits, 14
divisibility criteria, 17
divisor, 10

equation, 35
even integers, 10
exponential functions, 65

factorial, 104
function, 39

gradient, 44

inequalities, 38
inequality, 24
integers, 10
intercept form, 44
interest, 66
irrationals, 13

mean, 113

median, 113
mode, 114
multiple, 10

natural numbers, 9
negative numbers, 9
normal curve, 118

parabola, 51
percentages, 30
permutation, 103
polynomial, 37
prime number, 10
probability, 108
Pythagorean theorem, 71

quadratic equation, 51

radians, 43, 91
range, 28
rate problem, 99

ratio, 32
rationals, 11
rectangle, 81
rectangular prrism, 85
rhombus, 81
right triangles, special, 74

scatter plot, 120
sequence, 27
slope, 44
squares, 17
standard deviation, 114
system of equations, 47

trapezoid, 82

variables, 35
vertex formula, 54

weighed average, 113

www.ingramcontent.com/pod-product-compliance
Lightning Source LLC
Chambersburg PA
CBHW080006180726
48002CB00021B/3127